SpringerBriefs in Applied Sciences and Technology

Safety Management

The *SpringerBriefs in Safety Management* present cutting-edge research results on the management of technological risks and decision-making in high-stakes settings.

Decision-making in high-hazard environments is often affected by uncertainty and ambiguity; it is characterized by trade-offs between multiple, competing objectives. Managers and regulators need conceptual tools to help them develop risk management strategies, establish appropriate compromises and justify their decisions in such ambiguous settings. This series weaves together insights from multiple scientific disciplines that shed light on these problems, including organization studies, psychology, sociology, economics, law and engineering. It explores novel topics related to safety management, anticipating operational challenges in high-hazard industries and the societal concerns associated with these activities.

These publications are by and for academics and practitioners (industry, regulators) in safety management and risk research. Relevant industry sectors include nuclear, offshore oil and gas, chemicals processing, aviation, railways, construction and healthcare. Some emphasis is placed on explaining concepts to a non-specialized audience, and the shorter format ensures a concentrated approach to the topics treated.

The *SpringerBriefs in Safety Management* series is coordinated by the Foundation for an Industrial Safety Culture (FonCSI), a public-interest research foundation based in Toulouse, France. The FonCSI funds research on industrial safety and the management of technological risks, identifies and highlights new ideas and innovative practices, and disseminates research results to all interested parties. The series does not accept unsolicited submissions. For more information: https://www.foncsi.org/

FONCSI

Foundation for an Industrial Safety Culture

Corinne Bieder · Caroline Kamaté · Hervé Laroche ·
Eric Marsden · Jean Pariès · René Amalberti
Editors

Living with the New Safety Landscape II

Rethinking Safety Management in the Context of Cascading Uncertainties

Springer

Editors
Corinne Bieder
FonCSI
Toulouse, France

Caroline Kamaté
FonCSI
Toulouse, France

Hervé Laroche
FonCSI
Toulouse, France

Eric Marsden
FonCSI
Toulouse, France

Jean Pariès
FonCSI
Toulouse, France

René Amalberti
FonCSI
Toulouse, France

ISSN 2191-530X ISSN 2191-5318 (electronic)
SpringerBriefs in Applied Sciences and Technology
ISSN 2520-8004 ISSN 2520-8012 (electronic)
SpringerBriefs in Safety Management
ISBN 978-3-032-29342-8 ISBN 978-3-032-29343-5 (eBook)
https://doi.org/10.1007/978-3-032-29343-5

This work was supported by Fondation pour une Culture de Sécurité Industrielle.

This Springer imprint is published by the registered company Springer Nature Switzerland AG
The registered company address is: Gewerbestrasse 11, 6330 Cham, Switzerland

Preface

Safety has made significant progress since the advent of high-risk industries. Major accidents are fortunately rare nowadays, and there are a number of clues indicating that it is more and more taken for granted. Yet, if the historical approach to safety has made these safety results possible, the recent evolution and transformations of the world, and those to come, are likely to challenge current safety models, approaches, and practices, beyond the need for minor adjustments at the time. Indeed, major changes are ongoing at an unprecedented pace. Some are massive; they change the whole world or parts of it in many respects including safety in the way it is perceived but also the way it can be maintained or enhanced. Some changes are more local and affect safety in a more direct and obvious way. This acceleration of the transformations affecting safety urges the safety community to think ahead and get prepared to face unprecedented situations and expectations in the future, including a relatively near one.

This book in two volumes addresses the topic of the future of safety management and governance from as wide a perspective as possible to shed light on the current situation, its destabilization in the face of the many changes underway and to come and the possible ways forward to live with the inevitable uncertainties, complexities, inconsistencies. Drawing up a global picture is the purpose of the first chapter of the first volume meant to be the backbone of the two volumes and authored by FonCSI experts. The other chapters develop more detailed analyses and proposals on selected aspects addressed by the backbone chapter to substantiate the global view. These chapters are authored by international experts of these topics based on their research work and experience. Volume 1 starts with the backbone chapter *The crumbling facades of industrial safety*, before addressing new frames for a changing world through seven complementary chapters. Volume 2 discusses how organizations confront safety challenges in a first part comprising seven chapters, then provides illustrations of destabilizations and ways forward through Cases and examples in a second part comprising four chapters, before standing back again and developing a wide reflection on the safety governance regimes in its final chapter *Regimes of fractured objectivity: Beyond the Envelope of Technology Safety Governance.*

The very recent changes especially in the political or geopolitical domain (e.g., dismantling of regulatory bodies or research labs) tend to make any reflection on the topic almost immediately outdated, at least partly. The content of this book is no exception. It already evolved from what it was at the beginning of this initiative and would probably not be exactly the same should the book be issued a few months later. Nevertheless, the aim is not to be completely accurate (who could pretend to be so when it comes to addressing the future), but rather to start a broad and collective reflection—being aware of the uncertainties and remaining modest—and trigger initiatives to keep up with an evolving world.

Toulouse, France

Corinne Bieder
Caroline Kamaté
Hervé Laroche
Eric Marsden
Jean Pariès
René Amalberti

Contents

Part I
Organizations and Safety Challenges

Chapter 1
Lost in Representation: How Social Interaction Contributes to Safety

Petter Almklov and Stian Antonsen

Abstract The role of "informal" dimensions of work and how they contrast with formal safety management systems has been widely described in safety science. This chapter aims to unpack the nature of such dimensions. We show how social interaction is a foundational—rather than secondary—component of organizational safety. While safety science often focuses on formal structures, rules, and "work as imagined," we discuss how everyday interactions shape how organizations function and how safety emerges in practice. Drawing on micro-sociology and structuration theory, we illustrate how repeated social interaction generates trust, interagency learning, and collaborative forms, as seen in a case study of co-located emergency call centers. We discuss how digitalization, standardization, and new organizational forms change the conditions for interaction, potentially fragmenting social relations while enabling new, distributed forms of connection. We argue that social interaction fundamentally shapes organizations and how safety is achieved, today and also in the future. However, given new organizational forms and new technologies, it will be found elsewhere and in other forms.

Keywords Social interaction · Informal organizational processes · Work as imagined versus work as done · Digitalization and standardization

1.1 Introduction

Much of our research within the safety field has been concerned with "informal" dimensions of work and how they contrast with organizational discourses of work in general and those in safety management systems, rules, and regulations in particular.

P. Almklov (✉)
Department of Sociology and Political Science, Norwegian University of Science and Technology, Trondheim, Norway
e-mail: petter.almklov@ntnu.no

S. Antonsen
Statnett, Oslo, Norway
e-mail: stian.antonsen@statnett.no

C. Bieder et al. (eds.), *Living with the New Safety Landscape II*,
SpringerBriefs in Safety Management, https://doi.org/10.1007/978-3-032-29343-5_1

Social phenomena are often referred to in terms such as culture, safety culture, invisible work, communities of practice, experiential and tacit knowledge. These phenomena are hard to formalize, yet important elements of "work as done," and important for safety. In this respect, safety research is a particular interesting field for *organizational* research, since such qualities can be a matter of life and death.

Looking into the future, which is the goal of this book, safety theorists are rightly concerned about several developments in the landscape of new threats. Our scope here is not on the emergence or changes in risk or threats, but on the organizations responsible for managing these risks, how their internal workings may change, and how we as safety researchers must change with it.

The term "informal" is not a characteristic of social phenomena like culture, invisible work, or communities of practice per se. Rather it alludes to the fact that they are hard to capture in formal descriptions and prescriptions of work. So, the basic question is one of the representations. Safety researchers are interested in accidents and what can prevent them, and the informal dimensions are situational, context-dependent, personal, and social. They elude standardization, measurement, and generally the discursive form needed to fit in systematic description and hence govern. We have elsewhere discussed how informal qualities, often associated with success outcomes in resilience engineering and HRO research, are challenged by neoliberal organizing (Almklov and Antonsen 2010, 2014), digitalization (Almklov and Antonsen 2019), and tendering regulation (Størkersen et al. 2024). These are trends and developments that tend to be rationalistic in the sense that weight is put on work as standardized, episodic, discrete, and well-described entities. Work is described and governed as consisting of a set of standardized entities such as (standardized) tasks and workflows. These dominant rationalistic notions of work are integral in several converging trends such as standardization, the audit explosion, outsourcing and managerialism risk suppressing the fact that there is and, as long as humans are involved, always will be a representational gap between work as imagined (WAI)[1] and work as done (WAD) (e.g., Dekker 2006; Antonsen et al. 2008). In this chapter, we will move beyond the generalized recognition that the informal dimensions of organizations are there and that they are important and also can play a positive role for safety (as well as the negative in the form of non-compliance and mistakes). We will argue they are more than a residual category, an already existing trait of organizing that is defined by its opposite (the formal systems). They are not only red-listed species faced by expansive systems of rationalization, but also productive elements of everyday work.

Specifically, we will focus on how social interaction among workers not only supplements structure but more fundamentally shapes organizations. We define "social interaction" as the process by which human beings engage in some kind of exchange of symbolic communication where they interpret the communication and actions of others and adjust their own communication and interaction to the

[1] While the word "imagined" might suggest to some that this is a cognitive entity of sorts, this shorthand encapsulates all the ways organizational systems prescribe and represent work, and the discourses of work more generally in the organization.

other party. In the process, meaning is constructed not only with reference to the singular interaction, but also as expectations toward future interaction (see Blumer 1969).

While there are many facets of work as done that tend to elude the descriptions of WAI, social interaction is in some ways the "atom" of the informal dimensions of organization we seek to explore. Moreover, it is a useful lens to discuss how current and future changes in technology and organizing may influence the future safety and resilience of organizations. Even in the age of "living with" rapid technological change, increasing uncertainty, and geopolitical stress, they constitute aspects of safety and security which we simply cannot live without.

This approach is partly inspired by micro-sociological studies (see e.g., Fine 2012) showing that social interaction produces, reproduces, or alters the social structures that constitute the presence of something wanted—social interaction is where culture and communities of practice come from and is hence the very source of "the informal." We will discuss some of our own recent research on it and frame our analysis in a prospective manner, seeking to see how recent and future developments may be affected by our observations. But before we look into the future, let us consider some examples from existing research.

1.2 Social Interaction: Have We Taken It Seriously Enough?

Organizational safety science is, understandably, oriented toward outputs or performance—the way organizational characteristics and processes contribute to safety or the lack of safety. Put differently, organizational characteristics and processes are mainly relevant for safety researchers and practitioners insofar as they are linked to the production of safety or the lack of safety. This is the key difference between the applied discipline of safety science and all the bordering "mother disciplines" from which safety science sometimes borrows its underlying concepts and theoretical frameworks (Haavik et al. this volume; Antonsen et al. 2025; Rosness et al. 2026).

There is, however, an analytical sacrifice involved in framing our empirical studies in this way. It presupposes that we know a priori where to draw the line between "safety relevant" interaction and "irrelevant" interaction. This line is not always possible to draw. As was argued by Antonsen (2009) in his study of safety culture on offshore supply vessels, studying the relationship between social phenomena and safety will often involve a two-pronged approach: First you need to understand the social phenomena and the context they take place in, and only then can you make inferences as to how this influence safety. Following this argument, empirical studies (and the following grounding of theoretical development) may look very different if social interaction is put at center stage of empirical efforts, compared to it being placed in the background as a contextual factor. In such an approach, cultures, communities of practice, or "informal traits" are not only treated as "already there"—it opens for consideration of how they came into existence in the first place and are continually recreated and changed through processes of social interaction.

In a recent paper (Almklov and Antonsen 2025) we had such an opportunity to study social structures in the making. We studied how a recent co-location of the call centers of the police, fire, and health services influenced interagency collaboration in Northern Norway. We were expecting that there might be some straightforward operational results, that the physical proximity directly influenced their task resolution, that talking face-to-face or looking at each other's screens during emergencies could increase shared situational awareness and so on, but that was not our main observation. The most striking evidence in our findings was the process by which everyday social interaction between operators from the different centrals fundamentally shaped the organization: It led to interagency learning and trust, new collaborative forms, and we saw how rudiments of a shared "culture" or community of practice emerged. The everyday conversations and interactions about work and other things, the reciprocal sharing of resources and knowledge, were "informal" in the sense that most of it was not included in formal procedures. But it was more than a residual factor to the formal system. It was more than coping with the gaps of a system of protocols and procedures. Our analysis showed that these interactions were the cornerstone and engine of the co-location project as such, including how the formal structures were enacted and further developed. The title of the paper, "From trivial to critical," suggests that their ways of solving critical challenges in emergencies could be traced back to trivial, informal social interactions that were not at all related to their tasks of responding to emergency calls. The operators socialized and interacted in breaks and meetings and there was a seamless continuity between interactions that were directly relevant for their primary tasks, and chats about everyday things—the weather, what you did last weekend, how busy it was at work et cetera.[2] Though these interactions produced foundations for better collaboration, most of them were seemingly not motivated by such instrumental concerns. Social interaction led to, and reproduced, norms of reciprocity and trust that on the one hand could mean inviting the others over for a cupcake on a Friday, but in other situations it could lead to the dispatch of a fire truck to support an ambulance, an invitation to a joint psychological debriefing session, or collaboration in developing joint procedures and routines. The fact that centrals from three different agencies were co-located in the same building, and not placed in the usual separate locations, did not directly change their way of working if we consider, rules, protocols and procedures. However, it afforded informal social interaction that led to gradual changes in the ways they collaborated, and operated individually. These "informal" processes eventually also led to changes in formal routines and structures. In this sense, structure was shaped by interaction, and not (only) the other way around. As such, we found what microsociologists (e.g., Fine 2012) have observed before, and which Giddens (1984) has suggested in his structuration theory, that the small group level interactions are not only a corollary to

[2] Note that even these examples aren't completely irrelevant for practice. For example, knowing that the police central is busy a specific day would influence the threshold for the others for contacting them in situations where they are not strictly necessary, for example.

structure or shaped by their formal surroundings, nor are they only conservative forces resisting change, but also the other way around—micro-level interactions influence the formal structures and how they are enacted in profound and lasting ways. They shape their organizational surroundings.

One of our take-aways from working with that paper is that though organizational safety theory recognizes the importance of informal social processes, our understanding of it lacks depth and our conceptualizations of it lack nuance.[3] Many give primacy to "work as imagined," the formal structures prescribing and describing work, and see informal processes either as something to be changed or a reason for failure. In the case of Resilience Engineering and partly HRO, informal processes and social interaction are given a more positive role, yet mainly as supporting the execution of "the real work," the formal tasks. Questions are rarely posed regarding where these informal traits come from. Understanding the emergence of collaborative social interactions needs to go beyond the work execution and see how not directly work-related social interaction is part of the picture.

So far, we have focused on the basics of social interaction and its relationship to research and practice on safety, without connecting it to the disruptively changing conditions for safety research and practice. In the following we will discuss selected examples of how the conditions for social interaction in safety-critical work might be changing and the implications of this.

1.3 Digitalization and New Organizational Forms

1.3.1 Digitalization Communication Through "Thin Descriptions"[4]

The case study described in the first part of the chapter was, by and large, a story about geographical proximity and the way physical space nudged trivial social interaction in directions that provided basis for critical organizational qualities and the way critical tasks were carried out. Digitalization is in many ways the opposite—it involves loosening the relationship between work, interaction, and the physical surroundings. How does this change the premises for the processes and outputs we found in the physical context of the emergency call centers?

Digitalization and datafication represent pervasive technological and organizational developments that lead to new questions about social interaction, since they sometimes change the form, media, and actors involved in interaction. The mere recognition of this premise implies that interactionist perspectives deserve more

---

[3] In our paper on call centrals, we had to resist the urge of premature conceptualization, of referring to a joint culture or a community of practice, of trust or of shared situational awareness.

[4] The term "thin descriptions" is inspired by Porter (2012) who contrasts simple, but mobile and "orderly," representation (typically numbers) to fuller, contextualized descriptions (the term alluding to its contrast, Geertz' thick descriptions).

attention and explicit study in safety research. Since the breakthrough of digital communication in high-risk systems, e.g., the introduction of integrated operations in the offshore petroleum industry, a selling point has been that "everyone can interact with everyone" irrespective of geographical location. However, in many ways "the media becomes the message": The mediation of interaction involves channels with different bandwidths and there is every reason to expect that this influences both the information and the relationship between the actors involved in the interaction. For instance, using digital applications to report telemetry values to operators of SCADA[5] systems provides precise data on precisely the parameters it is designed for. This allows for a "thin description" of the situation at hand—there is no room for other meaning content than the values themselves. In contrast, having telephone contact with the on-site operators allows for thicker situational descriptions of the situation at hand and dialogue that is more oriented toward sense-making than sensing (e.g., "what do you see/hear/smell?"), providing the operator with a richer basis for making inferences and interpretations of the situation at hand. The point here is, of course, not that we should avoid new technologies and stick to established ways of working. The point is that the design of the medium for interaction will influence which interaction is possible and routinely practiced. The development of sensors and applications for data transfer will probably (although not necessarily) lead to more data-centric than actor-centric interactions.

In a similar fashion, one reported concern of nurses and doctors during the implementation of a new integrated system in the health sector in our region is the fact that automated information flows may reduce the breadth of communication between different units, wherein nurses and doctors often share additional information about patients. Importantly, they *can* still make a phone call or stop by to share or ask for such information, if necessary, but the established routines in doing so to get a joint understanding of the patient are removed, and thus also the interpersonal relations that accompany them. In the more data-centric organization, the role of humans in information exchange is different, and established interactional practices, routines, and norms are changed. However, as we will discuss below, it does not mean that they disappear.

1.3.2 New Forms of Organizing, Destabilization, and Disembedding of Relations

In a previous paper (Almklov and Antonsen 2019), we outlined digitalization's symbiotic relationship with forms of governance that are based on standardization. In the digital, data-centric organization work is conceived as the execution of standardized tasks, and flows of standardized information regulate and monitor work. Hence, people are increasingly interchangeable entities. Your local IT-support guy can be replaced by an offshored helpdesk because the task is standardized and the

---

[5] Supervisory Control and Data Acquisition.

information needed is disembedded from the physical site. Similarly, the engineering support for a complicated task is a remote expert-center, not the people you interact with daily. In both cases, such arrangements may be more effective and also better in terms of quality and safety, but they also change the nature of social interaction in task execution. You interact with someone serving a role or a function, often without knowing anything about the person filling that role.

Maybe the most striking case of techno-organizational fragmentation of the social dimensions of organization is found in the algorithmic governance of food delivery workers (see Nilsen et al. 2022). Their work is governed by a faceless algorithm in an app. At face value, this kind of work is completely atomistic: They don't have colleagues or a boss, just the app telling them where to pick up and deliver food. This is an extreme case, indeed, but similar forms of algorithmic governance are found elsewhere in more traditional organizations as well. As argued by our colleague Røyrvik (2017), neoliberal forms of governance, of which digitalization currently is a key element, more generally erode the basis of durable social relations in the work setting. If he is right, there are fundamental challenges to be asked about how and where social interaction occurs and what its future role will be for supporting safe work execution. More than eroding social life altogether, this should be read as a disembedding of social elements out of the physical work context. However, "the social" might still be there, and it might find new forms and arenas (see Kadolkar et al. 2025), but it might be less connected to work. If one considers communities of practice, for example, the strength of these is the entwinement of the social and work related. Social interaction is interwoven with practice. This might change with the data-centric and neoliberal organizations, although there is no necessity in it disappearing altogether.

1.3.3 Are We Heading Toward Atomistic Organizations?

In contrast to our co-location example in the beginning of the chapter, developments such as those outlined above can indicate that there might be less room and time for social interaction with co-workers, at least of a face-to-face nature. An important outcome of such interactions is, as in our call-central case, entanglement of personal and work relations, and the development we have described may suggest that these will become less likely to emerge. For safety researchers this means that we must question how qualities such as trust, safety culture, and communities of practice might fare in new work situations. It is easy, when you think of the app-governed lonesome food delivery driver, to imagine that the social dimensions of organization disappear or become irrelevant, and in some cases it might, but there are also other possibilities. This is the point where one can easily extrapolate trends and tendencies into bleak prophecies of a "brave new world of work" (Beck 2000) characterized by individualization, precariousness, and fragmentation.

However, this is not necessarily the case.If one postulates that many of the social dimensions discussed emerge based on repeated interaction and not necessarily depend on co-location,[6] one can see that also geographically distributed collaboration among co-workers can lead to reciprocal "cultures" with similarities to what we saw in the emergency call centrals. Wasko and Faraj (2005) observed what they call "electronic networks of practice" in which participants share knowledge and help other professionals based on social bonds and expectations, with many similarities to what we find in physical communities of practice. While these contributors may still need to be "bowling alone"[7] due to their physical distribution, their social capital is developed and resides in the network. Similarly, we see that peers find each other, and interact with each other, digitally or otherwise also in settings where co-location is not an option. Whether it can "replace" co-located interaction is an open, and empirical, question. It means, however, that we might need to look in new places for the repeated informal social interactions that we have suggested here is the basic foundation of informal social dimensions of organization. And if we, as can be expected, find some of them in digital chat groups, message boards, and the like, researchers and managers should remember that also such distributed social groups are emergent and informal.

Attempts to streamline, structure, and govern social interaction from above are likely to fail also in the digital future as much as it does today. Which forms social interaction may take in future organizations will be highly unpredictable and ultimately an empirical question. At this point of the argument, it should be clear that we are not proclaiming the death of social interaction as an aspect of safety in high-risk organizations or as an integral part of organizational life in any kind of organization. If digitalization or organizational restructuring eliminates arenas for social interaction, it is likely to re-emerge somewhere else and/or in new configurations. Figuratively, it seems that rationalistic organizing might be playing a game of whack-a-mole with informal social relations. If social interaction is suppressed at one arena, it might pop up at another arena and reappear in different shapes and forms in existing arenas that were not designed for the purpose. If researchers or practitioners wonder where it has gone, we should look for it. It might take place in the most unexpected of settings, and it can exert far more influence on the workings and outputs of organizations than it is usually credited.

1.4 Conclusion: Taking Social Interaction Seriously

Social interaction among humans is a key element of most workplaces. Most researchers also recognize its value for developing elements of practice that are invisible to and supplement the rationalistic descriptions of tasks as imagined. Through our examples and drawing on insights from micro-sociological theory and structuration

[6] See, e.g., Almklov et al. (2014) who challenge the "localist" tendencies of the literature on situated action.

[7] Referring to Putnam's (2000) classic "Bowling alone" on social capital. Their social networks are digital, so though they might bowl alone, they might play online computer games together.

theory, we argued that it is a fundament for as much as a supplement to organizational structure. Moreover, safety researchers, us included, typically focus on the products or sediments of social interaction, and not so much the processes themselves. When we talk about organizational culture and social culture, for example, it is in many ways an entification (Larsen 2010) of the sediments of repeated social interaction. We are drawn toward what is already there and disregard the processes whereby it came about. When organizations become digitalized and organized in ways that to a lesser extent afford physical encounters between colleagues, our empirical efforts should pay more attention to the productive side of social interaction. We need to understand the new ways that "informal" organizational qualities can develop, and the new forms they may take.

Inspecting social interaction in its own right, and on its own premises, allows us to study social institutions as they emerge, how day-to-day interactions contribute to the co-creation and shaping a community, networks, and cultural traits. This implies an interest in the verbs rather than nouns of organizing (Barley and Kunda 2001; citing Weick 1979; See also Almklov and Antonsen 2025). This interest has long traditions in organizational science and has found new expressions in studies of "organizational space," where researchers have moved to more process-oriented studies for understanding social interaction under different spatial circumstances and dynamics (e.g., Stephenson et al. 2020). This means tracing how repeated micro-level interactions shape their own surroundings. Repeated social interaction shapes its own conditions. The aggregate of previous exchanges shapes the expectations of the next one. If your colleague has helped you solve a problem before, you ask him for advice again. If you, for example, has learned to interact successfully with an expert that you have called before but never met, your previous interaction with him has sedimented into what could (but shouldn't) be called a "culture of two." What is formed by such repeated social interaction may be culture, or a community of practice, or good seamanship or something else, but these labels are post hoc and sometimes misleading. But, rather than only seeking to identify the result, one should pay interest in the process in which such informal relations develop. Within some strands of safety science, culture is often seen as from a psychological phenomenon, something "stored" in the heads of the involved agents, rather than in the relations in which it is actualized.[8] We also need to consider the interaction that precedes this result.

An important premise of this book is that both practitioners and researchers of risk find themselves in an "age of living with" a wide, varied, and growing menu of challenges that need to be tackled in order to uphold safety and security. While there are a lot of developments in the external threat picture to consider, this chapter has focused on how to develop and update our understanding of the resilience of organizations to live with techno-organizational change. The age of human beings interacting with other human beings is not about to come to an end any time soon, and high-risk organizations are no exemption to this. The technological, structural, and material contexts, the actors involved and the content of and medias for interaction

[8] A similar ontological point is also argued by Haavik's (2011) "From components to relations," there focusing on the relations between humans and technology.

change, but the processes and products of interaction still need to be properly understood in its own right, to understand the current role of "informal" organizational traits in upholding safety.

References

P.G. Almklov. S. Antonsen, The commoditization of societal safety. J. Contingencies Crisis Manag. **18**(3), (2010)

P.G. Almklov, T. Østerlie, T.K. Haavik, Situated with infrastructures: Interactivity and entanglement in sensor data interpretation. J. Assoc. Inf. Sys. **15**(5), 263–286 (2014)

P.G. Almklov, S. Antonsen, Making work invisible: New public management and operational work in critical infrastructure sectors. Public Adm. **92**(2), 477–492 (2014)

P.G. Almklov, S. Antonsen, Standardisation and digitalisation: changes in work as imagined and what this means for safety science, in *Safety Science Research* (CRC Press, 2019), pp. 3–19

P.G. Almklov, S. Antonsen, From trivial to critical. Emergent interagency collaboration through co-location of emergency call centrals. J. Contingencies Crisis Manag. **33**(1), e70015 (2025)

S. Antonsen, *Safety Culture: Theory, Method and Improvement* (Ashgate, 2009)

S. Antonsen, P. Almklov, J. Fenstad, Reducing the gap between procedures and practice lessons from a successful safety intervention. Saf. Sci. Monit. **12**(1), 1–16 (2008)

S. Antonsen, T. Aven, T.K. Haavik, Reflections on research in risk science and on risk science. *Risk Analysis: An Official Publication of the Society for Risk Analysis* (2025)

S.R. Barley, G. Kunda, Bringing work back in. Organ. Sci. **12**(1), 76–95 (2001)

U. Beck, *Brave New World of Work* (John Wiley and Sons, 2000)

H. Blumer. *Symbolic Interactionism: Perspective and Method*. (University of California Press, 1969)

S. Dekker, Resilience Engineering: Chronicling the emergence of confused consensus, in *Resilience Engineering: Concepts and precepts*. eds. by E. Hollnagel et al. (CRC Press, 2006)

G.A. Fine. *Tiny Publics: A Theory of Group Action and Culture* (Russell Sage Foundation, 2012)

A. Giddens, *The Constitution of Society: Outline of the Theory of Structuration*. (Univ of California Press, 1984)

T.K. Haavik, On components and relations in sociotechnical systems. J. Contingencies Crisis Manag. **19**(2), 99–109 (2011)

I. Kadolkar, S. Kepes, M. Subramony, Algorithmic management in the gig economy: a systematic review and research integration. J. Organ. Behav. **46**(7), 1057–1080 (2025)

T. Larsen, Acts of entification. The emergence of thinghood in social life. *Human Nature as Capacity. An Ethnographic Approach* (2010), pp. 154–176.

M. Nilsen, T. Kongsvik, P.G. Almklov, Splintered structures and workers without a workplace: How should safety science address the fragmentation of organizations? Saf. Sci. **148**, 105644 (2022)

T.M. Porter, Funny numbers. Cult. Unbound **4**(4), 585–598 (2012)

R.D. Putnam, Bowling alone. Simon and Schuster (2000)

R. Rosness, R. Flage, Antonsen, Viewing the risk concept as a boundary object: Implications for theory and practice. Risk Anal. **46**(6), (2026)

E. Røyrvik, Sosialitet i målstyringens tid, in *Trangen til å telle*, ed. by Larsen, Røyrvik (Scandinavian Academic Press) (2017)

K.A. Stephenson, A. Kuismin, L.L. Putnam, A. Sivunen, Process studies of organizational space. Acad. Manag. Ann. **14**(2), 797–827 (2020)

K.V. Størkersen, T.K. Haavik, P.G. Almklov, A.Å. Gauteplass, S.H. Jore, Unprocurable essentialities: Situational and relational knowledge in publicly procured security services. Saf. Sci. **178** (2024)

M.M. Wasko, S. Faraj, Why should I share? Examining social capital and knowledge contribution in electronic networks of practice. MIS Q. 35–57 (2005)

K.E. Weick, *The Social Psychology of Organizing*, 2nd edn. Reading. (Addison-Wesley, 1979)

Chapter 2
Living Safely with an Evolving Operational Environment

Tony Licu

Abstract This paper reflects on the dual perspective of a safety manager turned operational leader in air traffic management. It highlights how safety has gradually lost visibility in Ultra-safe Systems, constrained by limited budgets, outdated Safety-I thinking, and a lack of practical tools. Operational realities—political pressure, delivery deadlines, and competing objectives—create continuous trade-offs where safety risks are being sidelined. Emerging challenges such as geopolitical instability, cybersecurity threats, digitalization, AI, and talent shortages further strain safety management. The paper proposes modernizing safety governance, integrating safety professionals into operational teams, leveraging AI for proactive analysis, strengthening education for engineers and managers, and nurturing a culture of transparency to enable more effective, resilient safety practices.

Keywords Safety-I · Safety-II · Safety management · Security management · Human factors · Trade-offs · High-risk industries

2.1 Introduction

What is safety? Is it the first and foremost priority that should receive everyone's attention? Is it just one objective among many others? Or is it even a constraint in business development projects? The answer is a matter of perspective. Safety managers would probably share the first viewpoint, whereas operational managers might recognize themselves more easily in the second. But what happens when the same individual changes jobs from one managerial position to the other?

This chapter presents the experience of Tony Licu who spent his whole career in a high-risk sector, namely air traffic management. He was safety manager in a division doing applied safety research, development, training, and operational safety and human factors support for decades, from 1999 to 2024. Toward the end of this period, he was concomitantly head of a department focusing on digital transformation. In

T. Licu (✉)
EUROCONTROL, Brétigny-Sur-Orge, France
e-mail: antonio.licu@eurocontrol.int

C. Bieder et al. (eds.), *Living with the New Safety Landscape II*,
SpringerBriefs in Safety Management, https://doi.org/10.1007/978-3-032-29343-5_2

2024, he changed position to become an operational manager (Head of Technology) in charge of the development of a one of the biggest projects in EUROCONTROL's history, to modernize and digitalize the Agency's air traffic management assets infrastructure—the integrated Network Management Programme iNM. This also involved a transition from managing around 30 staff to managing around 700 staff and contractors, with a budget increase that is hard to imagine for most safety managers. Less than a year after this shift in his career path, this dual experience leads to several reflections on safety management as it is today and how it could evolve in the future, from these rare dual perspectives.

The first section focuses on his diagnosis of the current situation in terms of safety management from both perspectives. The second section presents his views in terms of ways forward regarding safety management and governance. Having been successively a safety manager and then an operational manager constitutes a particularly interesting experience for anticipating the foreseeable impact of current changes in the world on safety management. Indeed, this experience allows one to feel from two symmetrical, and sometimes opposing, points of view the fact that safety always results from trade-offs between several risks, constraints, and performance objectives, which are partly contradictory. And precisely, the changes underway will considerably modify the parameters of these trade-offs and their relative hierarchy and will make them increasingly complex and potentially unfavorable to safety.

2.2 A Dual Diagnosis of Safety Management Today

2.2.1 A Safety Manager's Perspective

2.2.1.1 A Downgrading of Safety in the List of Business Objectives in Ultra-safe Systems

In aviation, like in other high-risk industries, safety has long been promoted as the first priority. But with the absence of major accidents, this "golden age" appears to be over. Not only did the industry move away from stating safety as the first priority but simply securing sufficient resources—budget and people—is becoming difficult. When accidents are extremely rare, everyone, including decision-makers, considers that the activity is safe, making it hard to justify costs or investments, and even worse, leading to the idea that certain cuts can be made. Training and safety resources typically seem easiest to cut.

But past successes don't guarantee future ones. As with every operational manager, as a safety manager you have to repeat them. It's an educational exercise to get the message across on why you need to invest in safety, when budgets are under pressure. Education is key to achieving this, especially for managers. Managers have little training in safety. Those who undertake MBAs will typically find no safety content. It can therefore be seen as something for the safety department, unlike efficiency. So, we need to teach safety management in all management training, including MBAs. Likewise for engineers. The curricula in these training courses, while better, are only superficially covers safety around Europe. Some students can finish a Bachelor's degree in aerospace engineering without safety training.

At a wider scale, media play a key role in the perception of safety by the public. They only speak about safety when there is an accident. In safety, you can have a slow erosion on a daily basis. In air traffic management, there may be a slight increase in incidents or outages, but the evolution is slow and the numbers are low. The quandary is how to make the media talk about safety, and thus keep it as a lively concern, when there is no catastrophic event. Unlike most other goals, safety lacks salience in the public eye. Only obvious episodes of "unsafety" are typically reportable.

One objective that is becoming high on organizations' priority list is security. Security events involve ransomware and reputation attacks are becoming more and more numerous and powerful. It's no longer bright kids in their bedroom showing what they are capable of. Today, organizations are hit not only by activist group but also by States sponsored activities'. And that's a different scale. And so security is in the news every day, everywhere. In this context, more resources are naturally dedicated to security than safety. Security is more visible to managers and decision-makers and is a more salient threat to business continuity.

Yet, neither safety managers nor security managers have much of a voice in organizational decision-making. Safety managers have never been accepted with the same rank of seniority as other managers, especially compared to directors of operations and engineering. Safety managers are seen as the "poor cousin of the family," with a modest budget to help ensure a "dynamic non-event" (Weick 1987). The massive ratio that exists in terms of resources (budgets and people) between engineering or operations departments and safety or security departments somehow reflects the importance given to safety in organizations. Likewise, security managers are other "poor cousins of the family," even though more financial resources are allocated to security than to safety.

As another result of this declining interest in a somewhat taken-for-granted safety is that fewer and fewer people are willing to work in safety where salaries and visibility are less attractive than in other areas.

Despite the abundant literature addressing the dynamic construct of safety, or describing safety as an emerging property, in reality, organizations rarely give much prominence to "the safety people" because of the scarcity of accidents.

2.2.1.2 Safety Management: Advanced Thinking Hindered by Primitive Practical Tools

This perception about the level of safety is closely related to the safety model that most top managers implicitly have in mind, also called Safety-I (EUROCONTROL 2014a). This model looks at safety from a failure or deficit perspective, and relies on the number of accidents, incidents, and fatalities as safety indicators. These indicators are the lens through which managers and decision-makers understand and evaluate safety management. This approach to safety measured through its consequences dates back to the 1930s (Heinrich 1931) and inspired tools and methods that are still widely used today. With such models in mind, in the years following a major accident, it is much easier to get funds for safety, as observed between 2002 and 2008 [see Charles de Gaulle (HindSight 1), Milan Linate (2001), Überlingen (BFU—Investigation Report) accidents].

Until recently, this Safety-I model, based on failure to comply with the anticipated way of operating, was the only and default way to look at safety. Yet, since the 1980s, other approaches looking at how organizations are operating safely in context were suggested, such as HRO (2021) and normal accident theory (Perrow et al. 1999). Based on field observations, these theories and models propose a more realistic view on the organizational features that enable highly reliable organizations to manage safety effectively in real situations despite a high level of risk exposure. Later EUROCONTROL (2013) and Hollnagel (2014) coined the term "Safety-II" to designate the approaches to safety focusing on how things go well in the field, rather than how things go wrong. However, this type of approach has so far had a hard time making its way through in high-risk organizations. The main obstacle to the evolution of safety management toward this kind of model as claimed by managers and decision-makers is the lack of associated practical tools.

Safety management experts need to be honest with themselves. Sometimes they develop "white elephants" that cannot be implemented in daily activities. There is a credibility issue. Safety people often stay at the level of safety methodology or philosophy. A safety manager needs to be practical; otherwise, operational managers won't be interested. Translation is needed from theoretical insights and models to practice.

More recent approaches to safety stem from systems thinking accounting for the complexity or real situations, where multiple business objectives, including safety, co-exist. For such approaches, this translation effort is even more critical. Rather than counting the failures of safety through incidents and accidents, these approaches look more globally at how organizations deliver operations and performance, all safely. Such a perspective does not obviously translate into simple methods and KPIs. However, translating this sophisticated perspective on safety into practice can be done, as initially illustrated by the white paper written by EUROCONTROL (2014b). This requires understanding the perspective of operational managers, their context, objectives, and challenges. *When I was sitting just in safety, I was wondering how come these directors of operations or engineering don't understand or don't account for safety….*

2.2.2 An Operational Manager's Perspective

> *But when I moved to the other side, there is so much pressure to meet the deadlines – whether the pressure is political or financial – that you always wonder "what can I descope...what can I take out?".*

This experience of changing position from safety manager to operational manager makes you understand that in real-life, safety is different from theory. Operational managers constantly must adjust and do the correct trade-offs without compromising safety, despite the temptation to cut corners to relieve part of the pressure. That's more an art than a science (Shorrock 2014).

From this perspective, safety can be perceived as a nuisance. Safety specialists come to the program teams asking operational people to do "one more iteration" or to put together specific documents. In terms of governance, it was decided to separate safety from the other activities to be totally impartial. But with time, safety specialists lose operational *savoir faire.* While they are very good at safety processes, putting together safety cases, and of course verification, there is a disconnect from the other business objectives. Projects evolve without necessarily telling the safety people. And safety experts are not integrated into the project teams to help and advise them in real time. The operational manager then arrives at a decision to go ahead or not.

Key question operational managers have is: what added value do safety experts bring, not only in the investigation of incidents, but in day-to-day activity? The answer is a matter of person, background, experience, and reflexivity.

> *My strong background and experience in safety led me to bring the safety person who replaced me in the safety department together with the program team, to embed the safety guys into the program team. I try to convince the engineering and development teams to consult the safety experts before making a decision and anticipate the impacts a decision may have. As engineers or developers, they might not see all the impacts because they are working on their isolated topics, whereas safety experts have a helicopter view and look at the impacts of a decision from all the angles of the program.*

But this approach is strongly linked to a significant background in safety. Operational managers without such safety background would do things differently. It's not that they don't want to integrate safety, but they don't know about it and how to proceed in practice. It is critical to understand where they come from. Nobody wakes up and comes at work to kill people or to do a bad job.

Everyone wants to do well on safety, operations, and engineering. But there are some tensions between these objectives. This is where the art of trade-offs and compromises comes into play.

2.3 Upcoming or Growing Challenges

With the growing geopolitical tensions, from an air traffic management viewpoint, there are and will be many more airspace constraints, for example at the borders of some countries. Some airspaces are and will be closed. It induces increased complexity and density of traffic.

Security threats are increasing in number and severity. It is noteworthy that the aviation system was originally designed largely without protections, and for more stable, "peacetime operations." The future systems need to include more protection to cope with malicious interferences by States and organizations.

With digitalization, another major challenge is that of the power supply to run the data centers. But with the growing societal concern of climate change, these data centers need to be supplied with green energy, and hence in countries where the power is renewable. But data is often transferred using big underwater cables that are exposed to malicious acts. In the future, more resources (budget more than people) will need to be allocated to security, which will further increase today's difference between safety and security resources perhaps by a few orders of magnitude. Furthermore, there are insurances for safety events, even the most catastrophic ones like an air collision in the Air Traffic Management (ATM) domain. Will it still be the case for security? Will the legal system hold or evolve to get damages paid?

Artificial Intelligence—AI is another area of new safety challenges. One relates to certification as practiced in "classical" safety. Using machine learning in operational activities is new to the business as well as to the authorities. It's a hurdle to both parties that require new ways of working together. Another one concerns knowledge and sources of truth. Although the challenges posed by AI in the education and research worlds are widely acknowledged (e.g., some scientific papers are written by AI, with associated measures to detect it), the use of AI raises the question of the trustworthiness of the tools and methods that are used in management of safety. Safety analyses and safety cases require knowledge. With the decrease in competent resources will these safety approaches still be based on actual knowledge?

Attracting and keeping competent people with "a sense of safety" are becoming harder, especially concerning software developers. With a focus on the number of lines of code, the safety and security of the code is secondary: efficiency over thoroughness (Hollnagel 2009). When they enter an industrial sector such as aviation, they need to be trained in safety. But often, they leave to join IT companies because of financial and lifestyle incentives. Today, and even more so tomorrow, a few giant technology companies determine how things work for the rest of the world.

2.4 Possible Ways Forward for Managing and Governing Safety

2.4.1 New Opportunities to Manage Safety Differently

AI and in particular machine learning are offering new opportunities for safety to do analytics with normal operations, reflecting the focus of Safety-II. It's an opportunity to put buzz words into practice and identify weak signals by processing this massive amount of data on day-to-day operations. This is only becoming possible with today's technology. These opportunities might become an incentive to think differently about safety.

2.4.2 Developing a Systematic Safety (and Security) Questioning Reflex Through Education

There is a huge education challenge for developers, engineers, computer scientists, and managers to think about safety and security. Teaching safety (and security) management to students in aerospace engineering, and management, even at bachelor level, should be the norm.

Likewise, it is critical to teach in all the MBAs the principles of safety management and security management and the need to look at these with real eyes, not only through budget and accounting and KPIs. This also involves revisiting the current safety Key Performance Indicators (KPIs), which tend to be negative and have to be minimized like the number of accidents. Managers want to maximize the performance and thus the KPIs. But teaching this takes time. At least a week should be dedicated to safety and security management in Master of Business Administrations (MBAs).

2.4.3 Revisiting the Current Governance Regime

In the name of independence, the current governance of safety involves two splits:

- between regulators and operators,
- between safety and operations within companies/organizations.

What we observe today is that the split between regulators and operators or industrial organizations ensures independence, but regulators lose track of the content. Likewise, the split between safety and operations within companies ensures impartiality but leads safety people losing sight of the rest of the business.

These splits might need to be revisited to work more closely. If we keep the independence, we need to give the same incentives, financially, to the regulators and operators. Otherwise, there will always be poor relatives. By bringing safety and business together, we would help both safety, engineering, and operations. In the current situation, it is hard to have a genuine conversation between operational and safety people. Yet, being transparent about the pressures is needed. The challenges need to be discussed openly.

2.4.4 Nurturing a Company Culture of Transparency and Openness on Business Tensions

Besides the governance regime, it is also a matter of company culture. Sometimes, operational managers are under so much pressure that they don't have time to "think safety" and do the additional iterations asked by the safety people. But a manager should be able to expose these different views and have an honest conversation with everyone, explaining the various pressures and the reasons behind them. When people make decisions related to safety, most of them claim that they do this because it is under regulation. But by so doing, you undermine the culture. The true reason should be that they do it because it is good for aviation, for everyone, and for the companies.

> *I think if you are smart enough to work on the culture of the organisation, you can do amazing things. As long as you have a genuine conversation with your guys then you can see miracles happening. It is not easy today to have that genuine conversation. The safety experts are the White Knights and the top leadership are the bad guys because they come from a different perspective. I had that view until I moved to an operational management position.*

Nurturing the culture can happen through with small actions like inviting people to come have breakfast together. These initiatives turn out to be very powerful and in easy reach of any organization.

References

BFU—Investigation Report: AX001-1-2/02 Uberlingen Mid-air. SKYbrary Aviation Safety. https://skybrary.aero/bookshelf/bfu-investigation-report-ax001-1-202-uberlingen-mid-air

EUROCONTROL, *From Safety-I to Safety-II: A White Paper* (2013) (2437.pdf on Skybrary)

EUROCONTROL, *Systems Thinking For Safety: Ten Principles A White Paper* (Eurocontrol, Brussels, 2014a). https://skybrary.aero/sites/default/files/bookshelf/2882.pdf

EUROCONTROL, *Systems Thinking for Safety: Ten Principles A White Paper Moving Towards Safety-II* (2014b)

H. W. Heinrich, *Industrial Accident Prevention: A Scientific Approach* (1931)

High Reliability Organization a Complete Guide—2021 Edition—The Art of Service. 9781867420361. (High Reliability Organization Publishing). https://www.amazon.co.uk/High-Reliability-Organization-Complete-Guide/dp/1867420368

HindSight 1—Runway Incursion—Collision between Shorts 3-30 and Boeing 757 at Paris CDG. SKYbrary Aviation Safety. https://skybrary.aero/bookshelf/hindsight-1-runway-incursion-collision-between-shorts-3-30-and-boeing-757-paris-cdg

E. Hollnagel, *The ETTO Principle: Efficiency-Thoroughness Trade-Off: Why Things That Go Right Sometimes Go Wrong* (2009)

E. Hollnagel, *Safety-I and Safety-II: The Past and Future of Safety Management Paperback* (28 May 2014)

MD87/C525, Milan Linate, 2001. SKYbrary Aviation Safety. https://skybrary.aero/accidents-and-incidents/md87-c525-milan-linate-2001

C. Perrow, *Normal Accidents—Living with High-Risk Technologies* (Updated Edition Princeton Paperbacks, 1999)

S. Shorrock (2014). https://humanisticsystems.com/2014/10/20/if-it-werent-for-the-managers/

K.E. Weick, Organizational culture as a source of high reliability. Calif. Manage. Rev. **29**, 112–127 (1987)

Chapter 3
The Power of the Informal and Unspoken in Safety Practice

Jan Hayes and Sarah Maslen

Abstract Risk and safety are seen in most organizations as best managed by a combination of organizational structure, policies, standards, and processes working together to ensure that people are doing the right things. This codified view ignores important informal and unspoken practices that contribute to safe outcomes. The ability to imagine worst-case outcomes is critical in identifying developing problems early. Small faults and failures are key sources of knowledge about system health. Embodied knowledge developed from time in the field allows office-based engineers to make safe design choices. This chapter shows that engineering competence goes well beyond abstract reasoning and factual knowledge. In the face of megatrends requiring transformed safety practices, we ignore such soft skills at our peril.

Keywords Learning from accidents · Learning from cases · Incident reporting · Safety imagination · Embodied knowledge

3.1 Introduction

Driven by business norms and regulation, companies across the range of hazardous industries manage risk and safety based on internal control principles (Power 2007). In this mode of thinking about major accidents and how to prevent them, effective risk management is a matter of having the right combination of organizational structure, policies, standards, and processes working together to ensure that people are doing the right things.

Engineering knowledge has not always been so codified. People have been doing what we would now call engineering since ancient times with monuments such as pyramids and aqueducts providing visual reminders of the skills of these

J. Hayes (✉) · S. Maslen
RMIT University, Melbourne, VIC, Australia
e-mail: jan.hayes2@rmit.edu.au

S. Maslen
e-mail: sarah.maslen@rmit.edu.au

C. Bieder et al. (eds.), *Living with the New Safety Landscape II*,
SpringerBriefs in Safety Management, https://doi.org/10.1007/978-3-032-29343-5_3

ancient workers. These monuments were built based on collective shared experience, but since those times, engineers have outgrown their "artisanal origins to become humanity's steward[s] of technological progress" (Wisnioski and Downey 2012: 15). This has happened significantly due to professionalization. Until the mid-nineteenth century, engineering knowledge was largely gained from practical experience in the field via apprenticeship models of learning (Davis 2021; Zhang and Yang, 2020). The professionalization of engineering moved the focus of engineering learning and competence to more codified forms of knowledge via the introduction of tertiary qualifications in parallel with the production of written codes of practice and standards to make work outputs more uniform (Lloyd 2009). While such strategies have been important in improving the reliability of engineered systems, some authors suggest that this has been at the expense of important practical skills, i.e., engineering competence requires more than just codified knowledge (Ferguson 1994).

In the safety domain, the rise of the HSE professional in the last six decades has similarly led to a more codified and standardized approach to dealing with hazards and risks (Provan et al. 2017). Again, the reduction in industrial deaths and injuries suggests that this approach to safety based on internal control principles (Power 2007) has had benefits. Reliance on formal and written safety management systems has also led to the development of audit and conformity assessment, whose explicit aim is to confine practices to those that are codified and so destroy informal, unapproved practices.

In practice, engineers working in hazardous industry make high-stakes decisions every day. Formal processes play an important role in guiding such choices, but professional judgment also relies on informal and unspoken ways of knowing and deciding. Based on our work over more than a decade, mainly with Australian pipeline engineers, we describe three important mechanisms that contribute to excellence in safety practice and yet are invisible when safety is only viewed as an internal control matter (Power 2007).

3.2 Imagining an Undesirable Future

It is more than two decades since Pidgeon and O'Leary (2000) first highlighted the importance of a safety imagination, i.e., an ability for workers in hazardous industries to link day-to-day activities with the potential for catastrophic outcomes. In our view, this remains one of the most important informal aspects of safety practice. The record of major disasters shows vividly what can happen when people fail to link their work to the potential for disaster. PG&E's gas pipeline failed in San Bruno, California, in 2010 killing eight members of the public and destroying the surrounding suburb. The conditions that led to the pipeline rupture arose because of the failure by many people at PG&E to see how their actions could contribute to failure of their company's pipeline. This includes things as banal as poorly competed work permits for site work at the terminal upstream of the failure point that led to high downstream pressure, and dummy data that was never replaced added to the pipeline geographic information system, contributing to poor pipeline integrity management decisions (Hayes and Hopkins 2014).

In extreme cases, we see this failure to connect actions to outcomes even when a disaster has begun to unfold. Enbridge's liquids pipeline failure in Michigan, also in 2010, led to the largest onshore oil spill in US history. The size of the spill was largely the result of a failure by control room personnel to interpret the numerous alarms they were seeing and hearing as a sign that the pipeline had failed. They continued to run shipping pumps for several hours after the rupture occurred, confused as to why no oil was arriving at the pipeline's destination. In fact, the line had failed and the oil was being pumped into the Kalamazoo River (Hayes and Hopkins 2014). Their ability to make sense of the data in front of them was hampered by their failure to imagine that such a failure had occurred.

In contrast to these examples of lack of imagination, we see that in many cases engineers love disaster stories for what they can learn and imagination plays a key part in that (Maslen and Hayes 2020). Indeed, for engineers, disaster cases are "stories with a message" (Herreid 1997). In making sense of accident cases in a training environment, we have seen engineers put themselves in the shoes of their professional colleagues (Maslen and Hayes 2020). Imagining themselves at work making typical engineering decisions knowing that the choices made at the time led to disaster is a powerful way to strengthen their safety imagination.

Engineers use their stock of remembered accident cases in at least two ways. Firstly, when trying to influence others regarding the need for action on safety, engineers draw on accident cases to show what can happen when a pipeline fails and the human cost that could result (Hayes et al. 2025). Senior professional engineers have told us this is a successful strategy with senior management when used sparingly (Maslen et al. 2021). Another way in which imagining worst cases contributes to engineering judgment is in making detailed technical choices between options: how safe is safe enough? During our research, several engineers have described in detail how they create personal scenarios to test whether a proposed course of action feels right. Examples include imagining taking their children into an industrial space (Hayes and Maslen 2015) or imagining their house is near a pipeline (Hayes et al. 2025).

3.3 Learning from Small Faults and Failures

Engineers are interested in hearing about major failures, but we have found their use of failure cases linked to safety decision-making goes much further. In addition to consideration of major disasters, small anomalies, faults and errors also embody a "gift of failure" (Carroll and Fahlbruch 2011). Rather than being fixed and then forgotten, each fault is a potential source of valuable insights into system health. As we have written elsewhere (Hayes and Maslen 2023), success and failure can be two sides of the same coin. Any failure has the potential to provide useful insights into the state of the entire system and so should be valued.

Indeed, engineers find failures compelling. As Petroski explains, "engineering understanding did not progress by looking at successes; it did so by looking at failures" and further "understanding comes from failure; success comes from understanding failure and acting upon this knowledge" (2012, p. 87). Our research into this area has shown that the value of past failures is not only awareness of the potential for disaster, but also a specific technical understanding that directly informs professional practice (Hayes and Maslen 2023). An engineer's store of cases cumulates over time and such knowledge is a key aspect of engineering expertise.

It is one thing for individual experts to have a memory of failure cases to draw on. Maintaining corporate memory of small failures is also seen to facilitate ongoing organizational learning about the state of the system and the need for improvement. For this reason, companies and industries as a whole put substantial effort into gathering information about past small failures and their causes in an attempt to learn how to prevent more serious events from occurring (Maslen and Hayes 2016). High reliability researchers working in the aftermath of the Three Mile Island accident investigated why some companies that operate complex technologies appear to have a better than average safety record. They found that such organizations are "preoccupied with failure," i.e., they pay great attention to the small things that go wrong and see them as a window into the broader performance of organizational processes and practices (Weick et al. 1999). They found that highly reliable organizations actively seek out opportunities to learn about how their safety performance might be improved.

Industries as varied as air traffic control, oil and gas production, and nuclear power generation have actively taken up the idea of learning from small failures, which has led to the proliferation of formal systems for recording and classifying incident reports. Hecker argued that such databases serve as "boundary-spanning objects" that transfer, translate, transform, and distribute knowledge across organizational and professional boundaries. They have become "points of connection within a widely distributed memory system, spanning the project, the individuals … the organization and the professional communities involved" (Hecker 2012, p. 433). In practice, however, formal systems for collecting knowledge have a significant informal component in providing a mechanism for disparate pieces of knowledge to be connected. This is particularly important for identifying trends in the record of small failures brought about by system flaws or weaknesses. Another way such systems are used informally is when specific pieces of knowledge are shared. Well-designed systems are widely used by engineers to seek out relevant information in the context of their day-to-day work; for example, to check whether a particular item has proved to be reliable before specifying it in a new use. Each of these informal mechanisms is important in the context of collective knowledge for preventing rare events.

3.4 Codified Versus Embodied Knowledge in Engineering

When we think about expertise, it is important to keep in mind different forms of knowledge and how they relate. Often when we think about engineering knowledge, it is the technical aspects of the role that come to mind reflected in codified knowledge in industry standards and organizational procedures. While this is a critical foundation to engineering practice and the operation of engineering-intensive organizations, these ways of knowing do not completely resolve the questions that engineers face in the course of their work.

Where in many work domains, practitioners readily recognize and express uncodified knowledge in terms of their embodied skills (for example, see Maslen 2025), engineers often avoid such language. That is, they do not freely talk about how engineering calls for a particular way of seeing, as would someone involved in medical diagnostic work (Carboni et al. 2023; Maslen 2017). However, engineers *do* readily express their connection to the field as part of their decision-making processes. On exploring the value of field experience, we saw that engineers understood it to work on their sense of what the technology that they are designing looks, feels, and sounds like in practice, and so what this means for construction and operation, and the management of risk (Maslen and Hayes 2022). They cultivate a critical embodied sense of the technology that they take from the field back into their office-based work.

This embodied understanding operates at many levels. First, this knowledge is critical to design decisions. A classic example is the control room at Three Mile Island where draft beer taps of different brands had been attached to switches on the control room instrument panel used for raising and lowering control rods to compensate for the poor design decisions that resulted in switches being indistinguishable—not a good design outcome for safe operations of a nuclear power plant! Engineers commonly reflect that they did not grasp the complexity or scale of their designs before spending time in the field.

Second, embodied understanding supports office-based engineers to appreciate the people involved in physically operating the asset and how this should be felt in their designs, the procedures that they play a part in writing, and so on. People are tangibly absent from design drawings and yet they can necessarily work in close proximity to hazardous infrastructure. In having a connection to the field, engineers become aware of the workers involved in constructing and operating assets and can then imagine them as they go about drawing lines on a page. The archetypal example of this is whether it is physically feasible for a worker to construct an asset as drawn. Engineers also reported heightened awareness of areas of risk to workers through spending time in the field.

Embodied knowledge is also critical to engineers' reasoning over failure modes and the implications of these possible failure modes for their practice. Being in the field serves as a reminder of the hazardous nature of the material and the impact on workers and the public if there was a catastrophic failure. The engineers we spoke to on this issue noted that the technology can fail in ways that are unexpected—for instance, through protections that have degraded over time—where an engineer is focused only on the theory or a calculation.

3.5 Conclusion

Engineering competence is typically grounded in considerations of abstract reasoning and factual knowledge. In this chapter, we have set out how other forms of thinking and reasoning influence engineering practice for the better. Imagining bad outcomes, sharing stories of small faults and failures, and making use of embodied knowledge are all key skills that we see engineers using every day although there are limitations to these ways of knowing too. A challenge with embodied understanding in particular relates to the nature of rare events. In short, there are many experiences that practitioners simply would not encounter in order to integrate these understandings into their office-based work. For this reason, other informal methods for building understanding, such as the sharing of stories, take on critical importance in the process of drawing together informative cases and rules of thumb to supplement the codified knowledge that structures practice. When traditional safety management practices are threatened by multiple megatrends, which call for either "coping with" evolution to current practices or even for more radical transformations, "soft" skills linked to informal and embodied knowledge must be considered. We ignore such ways of knowing that play a key, but under acknowledged, role today at our peril.

References

C. Carboni, R. Wehrens, R. van der Veen, A. de Bont, Eye for an AI: more-than-seeing, fauxtomation, and the enactment of uncertain data in digital pathology. Social Stud. Sci. **53**(5), 712–737 (2023). https://doi.org/10.1177/03063127231167589

J.S. Carroll, B. Fahlbruch, "The gift of failure: New approaches to analyzing and learning from events and near-misses". Honoring the contributions of Bernhard Wilpert. Saf. Sci. **49**, 1–4 (2011)

M. Davis, *Engineering as a Global Profession: Technical and Ethical Standards* (Rowman & Littlefield, 2021)

E.S. Ferguson, *Engineering and the Mind's Eye* (The MIT Press, 1994)

J. Hayes, S. Maslen, Knowing stories that matter: learning for effective safety decision-making. J. Risk Res. **18**(6), 714–726 (2015). https://doi.org/10.1080/13669877.2014.910690

J. Hayes, S. Maslen, Preventing major disasters: success and failure as two sides of the same coin, in *Routledge International Handbook of Failure*. ed. by A. Mica, M. Pawlak, A. Horolets, P. Kubicki (Routledge, 2023). https://doi.org/10.4324/9780429355950-24

J. Hayes, S. Maslen, S. Holdsworth, O. Sandri, The part that we play: engineers' perceptions of their responsibility for public safety. Saf. Sci. **187**, 106862 (2025). https://doi.org/10.1016/j.ssci.2025.106862

J. Hayes, A. Hopkins, *Nightmare Pipeline Failures: Fantasy Planning, Black Swans and Integrity Management* (CCH, 2014)

A. Hecker, Knowledge beyond the individual? Making sense of a notion of collective knowledge in organization theory. Organ. Stud. **33**, 423–445 (2012)

C.F. Herreid, What is a case? Bringing to science education the established teaching tool of law and medicine. J. Coll. Sci. Teach. **27**(2), 92–94 (1997)
B.E. Lloyd, A short history of professional engineers in Australia (2009). https://www.profengaust.org.au/articles/short-history-professional-engineers-australia/. Retrieved from 3 November 2021
S. Maslen, Layers of sense: the sensory work of diagnostic sensemaking in digital health. 3: 1–9. Digital Health **3**, 1–9 (2017)
S. Maslen, *Learning to Hear: The Auditory Bases of Excellence in Practicing Medicine, Climbing Mountains, Making Music, and Communicating in Morse Code* (Columbia University Press, 2025)
S. Maslen, J. Hayes, Preventing black swans: incident reporting systems as collective knowledge management. J. Risk Res. **19**(10), 1246–1260 (2016)
S. Maslen, J. Hayes, Case based learning among practicing engineers: design, facilitation and lessons learned. Cogn. Technol. Work **22**, 307–319 (2020). https://doi.org/10.1007/s10111-019-00569-0
S. Maslen, J. Hayes, "It's the Seeing and Feeling": How embodied and conceptual knowledges relate in pipeline engineering work. Qual. Sociol. **45**(4), 593–616 (2022). https://doi.org/10.1007/s11133-022-09520-8
S. Maslen, J. Hayes, S. Holdsworth, O. Sandri, When ethics is a technical matter: engineers' strategic appeal to ethical considerations in advocating for system integrity. Sci. Eng. Ethics **27**(4), 46 (2021). https://doi.org/10.1007/s11948-021-00324-7
H. Petroski, *To Forgive Design: Understanding Failure* (The Belknap Press of Harvard University Press, 2012)
N. Pidgeon, M. O'Leary, Man-made disasters: why technology and organizations (sometimes) fail. Saf. Sci. **34**, 15–30 (2000)
M. Power, *Organized Uncertainty: Designing a World of Risk Management* (Oxford University Press, 2007)
D.J. Provan, S.W.A. Dekker, A.J. Rae, Bureaucracy, influence and beliefs: a literature review of the factors shaping the role of a safety professional. Saf. Sci. **98**, 98–112 (2017). https://doi.org/10.1016/j.ssci.2017.06.006
K.E. Weick, K.M. Sutcliffe, D. Obstfeld, Organizing for high reliability: processes of collective mindfulness, in *Research in Organizational Behavior*, vol. 3, ed. by R.I. Sutton, B.M. Staw (JAI Press Inc, 1999)
M. Wisnioski, G. Downey, *Engineers for Change: Competing Visions of Technology in 1960s America* (MIT Press, 2012)
C. Zhang, J. Yang, *A History of Mechanical Engineering* (Springer, 2020). https://doi.org/10.1007/978-981-15-0833-2

Chapter 4
Hazards, Culture, and the Co-construction of Safety in Complex Organizations

Teemu Reiman

Abstract The chapter examines safety and culture, arguing that definitions of safety must be grounded in hazards while recognizing its socially constructed nature. The chapter differentiates between different types of hazards and inspects how they contribute to different types of cultures being formed. The chapter also looks at how culture for safety, and especially the concept of safety culture, has been used previously. Approaches that see culture as a tool are contrasted with those that see culture as a phenomenon to be understood. The chapter proposes that viewing culture as an emergent social process offers a way forward in safety management.

Keywords Organizational culture · Safety culture · Hazard typology · Sense-making · Socio-technical systems

4.1 Introduction

Safety can be defined as freedom from unacceptable risks created by *hazards*. Yet how this freedom is attained and maintained remains a critical question. In many organizations, safety has been implicitly understood as the absence of incidents, treating a lack of adverse events as proof of being safe. Such an understanding can become self-confirming: As long as nothing visible happens, the system is presumed to be under control. However, in complex socio-technical settings, safety depends on ongoing adaptation and anticipation that may only look stable if we pay attention solely to the outcomes (Weick 1987).

The new view on safety (Hollnagel 2014; Dekker 2015; Woods 2019) emphasizes that safety requires more than the absence of incidents. Safety emerges from what an organization does; it is the presence of certain practices, conditions, and capabilities. Enhancing such capabilities contributes not only to preventing harm but also to organizational performance. The new approaches to safety such as Safety-II, Safety

T. Reiman (✉)
Lilikoi, Lohja, Finland
e-mail: Teemu.Reiman@lilikoi.fi

C. Bieder et al. (eds.), *Living with the New Safety Landscape II*,
SpringerBriefs in Safety Management, https://doi.org/10.1007/978-3-032-29343-5_4

Differently, and Human and Organizational Performance (HOP) have broadened understanding of what is required for systems to be safe.

However, when safety is equated with general success, it risks losing sight of the specific hazards, trade-offs, and contradictions that give safety its distinctive meaning. A comprehensive definition of safety must combine the new view's focus on adaptive capability with the traditional concern for hazard control. Safety varies depending on the specific hazard, and we thus need to ask, "safe from what" and "safety for whom."

In organizational contexts, safety is an emergent property of what the organization does, as a byproduct of how the organization carries out its work. This means that an organization cannot control safety directly, but it can shape the socio-technical conditions from which safety emerges. Managing safety means managing the system and its capacity to handle hazards (Reiman 2026). The organizational core task, i.e., what the organization is trying to accomplish, is an inherent source of hazards for the organization. For some organizations, these hazards may lead to major accidents with consequences for people and the environment (e.g., nuclear power plants or offshore platforms). For others, the core task-related hazards may mostly threaten production with the potential for individual human harm (e.g., manufacturing equipment with injury potential).

To conclude, effective safety management requires grounding definitions of safety in the actual hazards people face, while also recognizing that safety is socially constructed through everyday practices and interpretations. In this sense, safety, both subjective and objective, is inherently context-dependent.

4.2 Objects and Subjects of Safety

We have established that safety is an emergent property of socio-technical systems and that its further specification requires a deeper understanding of the particular system in question, including the hazards it must manage. Understanding safety therefore requires asking both what we seek to be protected *from* (the *objects* of risk, that is, the hazards) and for *whom* protection is sought (the *subjects* of safety). Different types of hazards call for different management logics, while differing interests and perspectives determine which hazards receive attention and which are ignored.

We can differentiate between four types of hazards (Reiman 2026).

Core task-related hazards relate directly to an organization's core processes, technologies, and products. These can also be called system hazards, but as a system can be defined on several levels of society, I prefer the term core task-related hazards. This also differentiates them better from external hazards, which are sometimes called systemic hazards (or systemic risks). Managing core task-related hazards typically requires a centralized approach (cf. Hopkins 2019) where inputs from different parts of the system are gathered and integrated into a view of the current state of the system.

Local hazards arise from interactions between work conditions, methods, and people. They include falls, trips, electrocution, physical forces, chemicals, posture, temperatures, lighting, etc. These hazards vary between workgroups and situations, and thus the hazards can be managed best in a decentralized manner as the way these hazards should be managed depends on the current realities in the field. There are, however, potential links between the local and core task-related hazards, and these need to be coordinated in a centralized manner.

Psychosocial hazards concern employees' internal states such as stress, fatigue, and exhaustion, which are difficult to observe directly. Their physical manifestations can be visible though. Although internal, these hazards manifest through organizational mechanisms and are prone to external influences.

External hazards are hazards that originate from outside the organization. They can be dangerous themselves (e.g., floods, earthquakes) or exacerbate core task-related or local hazards (e.g., bad weather affecting landing conditions for an aircraft). The risks associated with these hazards are sometimes called "systemic risks."

Each type of hazard calls for a different mode of management. Core task hazards require a centralized, system-level coordination of information and expertise. Local hazards are best managed through decentralized practices that allow frontline adaptation and local control. Psychosocial hazards call for relational and participatory approaches that address wellbeing, workload, and interpersonal dynamics, often involving both management and HR functions. External hazards require strategic foresight and collaboration with actors beyond the organization's control. Differentiating among hazard types is therefore not merely analytical: It reveals why no single management mode can ensure safety. Effective safety management must balance these logics and connect local adaptation, system control, and strategic anticipation (Reiman 2026).

Because culture forms around the organization's core task, it inevitably reflects the specific hazards employees must manage, and the relative emphasis placed on each. Changes in the environment also influence existing hazards and give rise to new ones (e.g., climate change, AI). The culture of the organization plays a major role in how such emerging risks are recognized, monitored, and addressed.

Safety is never neutral: It serves particular groups and interests. Decisions about which hazards to control, what risks to accept, and what performance to prioritize reflect implicit judgments about *whose* safety matters most. In some contexts, the primary concern is the safety of employees; in others, it is the safety of the public, or the protection of the organization's continuity and reputation.

These orientations can coexist yet also compete, shaping how trade-offs are made and justified. Asking *safety for whom* reveals the value choices embedded in safety management and how culture distributes attention and responsibility across groups. It reminds us that defining and pursuing safety are not only a technical or organizational act but also a moral and political one.

These value choices also shape the cultural logics that define what the organization perceives as safe to do and from what it seeks protection. A cultural approach provides the lens through which these judgments of "safe from what" and "safe for whom" are made and justified.

4.3 Safety, Culture, and Safety Culture

Organizations create a culture when they solve their problems related to external adaptation and internal integration (Schein 2017). While doing this, employees learn practices and form beliefs and assumptions around the questions of what they are safe to do, and from what they are safe.

All groups and organizations develop their own culture. Culture is a proven way of acting, thinking, and feeling that is shared by a group or an organization. It provides a tacit frame that shapes what people attend to, how they interpret signals, and what they consider normal or acceptable.

As previously noted, culture forms around managing certain types of hazards. In other words, cultures *emerge* where meaning is constructed around work and its hazards. As a consequence, the organization may be good at dealing with certain hazards but less so with other types of hazards. What matters, therefore, is not only understanding the organization's culture as such, but how meanings about safety and risk are created and maintained in its social environment. Because different parts of the organization confront different types of hazards, they also develop distinct ways of making sense of and managing them.

Although organizations are often seen as having a single, shared culture, in practice cultures are fragmented across groups and professional communities. Each group develops its own way of making sense of the hazards it faces, reflecting differing assumptions about acceptable risk, authority, and expertise. For example, technical specialists dealing with core task hazards may value precision and procedural control, while frontline teams managing local hazards rely on adaptability and trust. Management groups addressing psychosocial or external hazards may emphasize strategic foresight and external support. Organizational culture is therefore not a uniform entity but a loosely coupled system of interacting cultural logics. Managing safety thus involves maintaining productive dialogue across these partially autonomous cultures, rather than enforcing a single coherent view.

The concept of *safety culture* has come to serve simultaneously as an explanation of safety problems and a remedy for them (Silbey 2009). This dual role has blurred the distinction between describing how organizations function and prescribing how they should function, reinforcing the need for a more integrated perspective (Reiman 2026).

There is no consensus on the exact definition of safety culture. It is variously understood as a management tool to influence behavior, as a causal explanation for events, as shared assumptions about safety, or as a description of how the organization constructs meaning around safety and its other goals (Guldenmund 2000; Silbey 2009; Reiman 2026). Each approach rests on distinct assumptions about causality and the nature of reality (see Table 4.1). While these differences can create apparent contradictions, they also offer complementary insights that together provide a richer understanding of safety culture.

When culture is seen as a remedy (i.e., as a tool), the focus is often on individual behavior which is seen as the problem to be corrected. In this view, the problems are

Table 4.1 Four different conceptualizations of safety culture based on Reiman (2026)

	Remedy	Causal explanation	Dynamic process	Emergent structure
Philosophical underpinning	Functionalist	Functionalist	Interpretive	Interpretive
Purpose of the concept	Tool to *improve* safety by changing behavior	Tool to *explain* why positive or negative safety outcomes occur	Phenomenon *describing* how safety is constructed and enacted by people	Phenomenon *describing* underlying logics of action in a group
Culture and behavior	Culture *is* behavior	Culture *affects* behavior	Culture is *in* the behavior, as practices	Culture is *beneath* behavior
Culture and safety	Correct behavior is safety	Culture directly affects safety	Culture and safety both emerge from interaction	Culture indirectly affects safety
Proof of usefulness	Does it help in *solving* already known problems	Does it help in *specifying* and *solving* problems	Does it help in *discovering* patterns	Does it help in asking more elaborate *questions*

known, and culture is the cure. It is based on a functionalist conception of groups and societies having a culture to bind them together.

The view of culture as a causal explanation is prevalent in many accident investigations and also quantitative survey studies (cf. Guldenmund 2000). The aim is to explain, and the approach is based on the premise that the cultural dimensions have a causal influence on behavior. The premise is, like in the remedy approach, that culture is functional and necessary for social order.

Seeing culture as a dynamic process highlights how everyday interactions continually recreate share meanings. Sense-making, as Weick et al. (2005) describe it, is the process of organizing through which people enact and interpret their reality, including hazards and safety itself. This is an important addition to the view on culture as it highlights the ongoing social construction of reality while acknowledging the existence of physical objects and hazards as one object of sense-making. The process view also highlights the self-referential nature of culture: We create culture for us but once created it "takes a life of its own" and affects how we interact. In this view, issues that receive less attention in functionalist theories such as power and conflicts (cf. Antonsen 2009) are also often addressed better.

Moving from a functionalist view of safety culture to a more interpretative view of culture as an emergent process and/or structure allows a more nuanced view of the organizational reality. It also indicates a shift from seeing culture as a tool to viewing it as a phenomenon, the context and outcome of behavior in organizations (cf. Silbey 2009).

The final perspective, culture as emergent structure, focuses on the layers of culture in an organization. The deepest layer of culture consists of shared basic assumptions and underlying logics of action (Schein 2017; White 2017). Basic assumptions are taken for granted, and it is difficult for members of culture to recognize or verbalize them themselves. These basic assumptions to a large degree define what is normal for a group, and culture thus influences what is considered normal work, what is success, and what is failure. To understand culture, it is important to map out these deeper contents. These assumptions affect how people interpret various internal and external signals and events. They create expectations and guide our attention.

Because interpretive views focus on meanings and sense-making, they may pay less attention to the material and causal dimensions of safety, such as concrete hazards, physical conditions, and individual behavior. Connecting interpretive approaches with hazard and accident models helps maintain the link between culture and the tangible realities of risk. More functionalist models, e.g., normative models of safety culture, can be utilized from an interpretive perspective as ideal models or frameworks that help to identify blindspots in the organization. In this way, culture can be seen as both process and structure, while functionalist models help highlight the key areas to address in developing a healthy safety culture (Reiman 2026).

The nature of the dominant hazards also influences which cultural logics take precedence. Organizations primarily exposed to core task hazards tend to value control, expertise, and centralized coordination, whereas those dominated by local hazards emphasize craftsmanship and local adaptation. Where psychosocial hazards are salient, cultures of care and inclusion become central, while external hazards call for strategic scanning, preparedness, and foresight. These emphases are not mutually exclusive, yet they compete for attention and resources. Over time, an organization's culture becomes attuned to the hazards it encounters most often and may downplay others, reflecting trade-offs and selective attention. Understanding which hazard types are culturally foregrounded or backgrounded is essential for interpreting any culture. The same pattern can be seen among safety professionals, whose expertise often centers on particular types of hazards, shaping what they attend to and how they define and manage safety.

4.4 Integrated Approach to Safety

Viewed from a cultural perspective, an integrated approach seeks to connect functionalist tools with interpretive insights, emphasizing sense-making and shared assumptions as levers for improvement. An integrated approach entails analyzing the interactions among various factors rather than addressing them individually. Power, culture, and leadership should not be viewed as isolated dimensions but as interdependent processes shaping system performance. Safety management should thus target the system as a whole instead of its individual components. Achieving such system-level understanding requires that interpretive approaches to safety management gain legitimacy and be recognized as contributing directly to the organizational core task. This issue also relates to the identity of safety professionals and the status of safety as a science (Le Coze 2024). To gain legitimacy, interpretive understanding needs to be translated into practical interventions by developing and implementing tools that make cultural and systemic insights more visible and actionable to management. This may require that safety professionals widen their interests and competence outside their choice of hazards.

Understanding safety within organizational context requires considering the overall system as greater than the sum of its parts. Often, in the wake of incidents, there is an effort to comprehend the situation by scrutinizing increasingly minute details. Although detailed analysis is important, it is even more crucial to appreciate the broader context and interconnections; a cultural perspective elevates our view beyond individual performance or single-point remedies.

In practical terms, emphasizing organizational capabilities integrates safety professionals more closely with quality and human resources functions, as well as with the actual conduct of work. Safety should be embedded in daily operations and viewed broadly as an organizational capability: the ability to conduct work successfully while maintaining freedom from various types of hazards. This capability rests on human, technical, and organizational capacities for proactive and reactive adaptation to changing conditions and emerging disturbances (Hollnagel 2014; Reiman 2026).

The next step for research and practice is to refine the interpretive understanding of safety and develop methods that make it usable and legitimate in everyday managerial work. This points toward what could be termed *actionable safety science;* research and practice that translate interpretive insight into legitimate managerial tools (Reiman and Viitanen 2019).

In an integrated approach, the focus should be equally on organizational structures, work practices, understanding of employees and collective mindset (values, norms, meanings, assumptions). Although safety development must focus on both abstract and tangible, it must be understood that all development of organizations and work is cultural, intentionally or unintentionally. The critical issue is ensuring that the development of practices, structures, and culture remains mutually aware and responsive, with mindfulness toward how safety and hazards are perceived and enacted in everyday work. Integration in this sense does not mean creating a single, uniform culture, but fostering dialogue and connectivity across different perspectives and domains that together shape organizational safety. Achieving integration calls for a shared shift in mindset across professional boundaries from safety specialists to HR, quality, and strategic management so that safety becomes an inherent part of how the organization thinks and acts, embedded in its cultural fabric.

As discussed earlier, different types of hazards call for different management logics; the challenge of integration lies in maintaining coherence across these modes. Thus, integration includes reconciling the different logics required by different types of hazards. Core task hazards invite standardization and system-level control; local hazards call for flexibility and local problem-solving; psychosocial hazards require sensitivity to wellbeing and workload; and external hazards demand strategic vigilance and capacity for rapid adaptation. An integrated approach recognizes that these logics coexist and must be balanced rather than allowing any single one to dominate.

Cultural maturity is reflected in the organization's ability to maintain dialogue and coordination among these differing hazard responses, ensuring that attention to one type of safety does not erode another. This ability has become even more critical amid today's turbulence, where digitalization, climate pressures, and geopolitical uncertainty blur hazard boundaries and challenge traditional safety tools. In such volatility, fragmented approaches are insufficient; safety depends on connecting diverse logics, coordinating across levels, and maintaining shared understanding. Integration, understood as connecting diverse people, technologies, and cultural logics, thus becomes a condition for organizational adaptability and resilience.

References

S. Antonsen, Safety culture and the issue of power. Saf. Sci. **47**(2), 183–191 (2009)
S.W.A. Dekker, *Safety Differently: Human Factors for a New Era*, 2nd edn. (CRC Press, 2015)
F.W. Guldenmund, The nature of safety culture: a review of theory and research. Saf. Sci. **34**(1–3), 215–257 (2000)
E. Hollnagel, *Safety-I and Safety-II: The Past and Future of Safety Management* (Ashgate, 2014)
A. Hopkins, *Organising for Safety: How Structure Creates Culture* (CCH, 2019)
J.-C. Le Coze, Understanding the 'blues of safety professionals.' Int. J. Occup. Saf. Ergon. **30**(2), 351–365 (2024)
T. Reiman, *Safety in Complex Organizations. A Cultural Approach* (CRC Press, 2026)
T. Reiman, K. Viitanen, Towards actionable safety science, in *Safety Science Research: Evolution, Challenges and New Directions*. ed. by J.-C. Le Coze (CRC Press, 2019)
E.H. Schein, *Organizational Culture and Leadership*, 5th edn. (Wiley, 2017)
S.S. Silbey, Taming Prometheus: Talk About Safety and Culture. Ann. Rev. Sociol. **3**, 341–369 (2009)
K.E. Weick, Organizational culture as a source of high reliability. Calif. Manage. Rev. **29**(2), 112–127 (1987)
K.E. Weick, K.M. Sutcliffe, D. Obstfeld, Organizing and the process of sensemaking. Organ. Sci. **16**(4), 409–421 (2005)
D.G.J. White, *Rethinking Culture. Embodied Cognition and the Origin of Culture in Organizations* (Routledge, 2017)
D.D. Woods, Essentials of resilience, revisited, in *Handbook on Resilience of Socio-Technical Systems*. ed. by M. Ruth, S. Goessling-Reisemann (Edward Elgar Publishing, 2019)

Chapter 5
Suffocation by Management Systems. Why Deregulatory Chainsawing Will Fail to Reduce Administrative Burdens

Kristine Størkersen and Sidney Dekker

Abstract This chapter explores the paradoxes of contemporary deregulation through the lens of safety management research. While some political narratives frame deregulation as a means to reduce bureaucracy and enhance efficiency and innovation, empirical studies have suggested that such reforms can create internal overregulation: Years of critical safety management research demonstrate how deregulatory measures like self-regulation and internal controls can generate excessive documentation, reduce operational discretion, obscure the quality work, and fragment accountability. Together, such research results show how deregulation has not given simplification and efficiency, but rather shifted burdens from the public to the private sphere. In this chapter, we explain how those studies give a warning about the future of both safety management and deregulation in general. Deregulation and overregulation are not opposites but co-occurring phenomena that give rise to new forms of bureaucratic complexity. This means that overregulation can be reduced not through blunt deregulation, but through deeper knowledge of the various demands of the private corporate environment. Through that understanding, we also can find future solutions for safety management, as well as regulation in general.

Keywords Deregulation · Overregulation · Safety management · Audits · Bureaucracy

---

K. Størkersen (✉)
Sintef, Norway
e-mail: kristine.storkersen@sintef.no

S. Dekker
Griffith University, Brisbane, Australia
e-mail: s.dekker@griffith.edu.au

C. Bieder et al. (eds.), *Living with the New Safety Landscape II*,
SpringerBriefs in Safety Management, https://doi.org/10.1007/978-3-032-29343-5_5

5.1 Deregulation Promoters Will Have to Eat Their Hats

Cutting regulations was meant to set companies free—but instead, it has left them tangled in even more complex management systems. Safety management research has revealed organizational rationalities that not only shape the future of safety practices but also may predict the unintended consequences of deregulation.

Political movements and business owners advocate deregulation to reduce costs for governments and companies. In recent years, many countries have witnessed aggressive "chainsawing" of governmental departments. This trend has intensified through right-wing initiatives such as the US and Argentina's chainsaw reforms, which pursue extreme regulatory slimming. A more common example of deregulation is laxer governmental control of companies through function-based rules designed to minimize regulatory and administrative burdens, in contrast to traditional "command-and-control" regulation with prescriptive rules.

Yet organizational research shows that deregulatory measures rather increase the regulatory load within companies. In the absence of prescriptive state rules, companies themselves develop internal bureaucracies centered on audits, documentation, and accountability mechanisms to manage liability and reputation (Dekker 2025; Størkersen et al. 2020). There is little evidence that companies will succeed in simplifying their management systems.

This chapter presents safety management research that explains why "chainsawing" regulations can backfire, as reduced state regulation is shown to trigger overregulation in companies. The next section provides background on deregulation and safety management research, followed by an analysis demonstrating why successful deregulation requires removing organizational traditions and auditism that currently drive internal bureaucracy.

All in all, this chapter shows how the future of safety management lies in the current organizational responses to function-based regulation and how this can illustrate the organizations' response to deregulatory measures in general. In the near future, organizations will need a total shift in, or more sophisticated, management tools to avoid getting suffocated by increased management demands.

5.2 Background

This section gives an introduction to regulation, deregulation, and how organizations react to deregulatory measures. Regulation is fundamentally about protecting societal values—such as worker safety, environmental sustainability, and public health—by limiting actors' discretionary space (Baldwin et al. 2011). There is a large research corpus on various forms of regulation and state, which is not addressed here, as this chapter is about the organizations' response.

5.2.1 Deregulatory Measures

Political discourses increasingly promote deregulation and state cuts as pathways to cost saving, efficiency, and innovation, although the empirical evidence is weak. Neoliberal programs have for decades stated that governments are too large, too bureaucratic, and too slow to adapt. A frequently mentioned early deregulatory reform was the New Public Management (NPM) promoted from the 1980s market-based governance, performance measurement, and privatization (Dekker 2025). Neoliberals thereafter reframed the state as a facilitator rather than a provider of services, saying governments should "steer, not row." There is also renewed interest in deregulation (Andrews et al. 2025).

Contemporary deregulatory initiatives center around the same aims and activities (see Table 5.1). All echo with earlier reforms, but some are more confrontational and extensive. Javier Milei's "chainsaw" metaphor in Argentina symbolizes cutting public spending and privatizes state functions. The 2025 DOGE initiative in the USA aims to "commence the deconstruction of the overbearing and burdensome administrative state" (White House Executive Order 14219, 2025), reducing public sector employment and canceling contracts. A much more common deregulatory measure is function-based rules, aiming to simplify regulatory burdens for government and companies. Even this mild deregulatory measure has consequences going against the intentions of simplification.

The organizational and management research behind Table 5.1 underlines that state deregulation has led to overregulation in companies. Companies meet their new responsibilities in ways that generate adverse effects, working against the primary intentions of deregulation.

5.2.2 Function-Based Regulation, an Omnipresent Deregulatory Measure

Two common types of regulation are *prescriptive* and *function-based* regulations. *Prescriptive* regulations do not offer much leeway for how to live up to legislated obligations. *Function-based* (or goal-based or performance-based) regulations, by contrast, set outcome-oriented goals such as safety or quality, leaving companies free to choose how to meet them (Lindøe et al. 2011).

Rather than government specifying how companies should operate, function-based rules require companies to implement and internally control management systems tailored to their operations (Lindøe et al. 2011). Relatively brief law texts state that companies are responsible for the consequences of their operations, while government only is to verify that they have and follow good plans. Auditing is the common tool to verify compliance, through verification of the organizations' internal control systems, management systems, or quality systems. This has been

Table 5.1 Aims of deregulation and examples of measures, scales, and results seen in European, American, and Australian governments and companies[1]

Aims of deregulation	Examples of deregulatory measures	Scale	Empirically seen results
Reducing government costs (decreasing the public sector)	Downsizing governmental departments; reducing budgets; privatization and outsourcing; minimizing public regulatory responsibilities	From radical/aggressive (removing entire public services) to common/mild (NPM; function-based regulation increasing private responsibility)	Increased government resources needed for enforcement
Reducing companies' costs (increasing innovation and growth)	Self-regulation; simplifying or eliminating rules and restrictions, particularly those not having apparent benefits ("red tape"); giving incentives and contracts to increase profit and innovation		Overregulation in companies: extensive management systems and internal bureaucracy

labeled self-regulation, but is in practice co-regulation, since companies and state regulate "together" through verification of internal control systems.

The last decades' shift from prescriptive to more function-based regulations is based in deregulation, as governments seek to reduce administrative burdens and encourage innovation while maintaining accountability and safety standards. While this may appear to simplify regulation, it has unexpectedly led to compensatory *overregulation* within private organizations.

Function-based regulation has led to the rise of internal control systems and bureaucratic complexity within organizations. In the absence of detailed rules, organizations must develop internal control systems to demonstrate accountability to not just regulators but also customers, insurers, and the public. This was described already in the 1990s (Power 1994). More recent empirical examples show ballooning overregulation in the private sphere, driven by their search for interpretation of verification requirements (Dekker 2025; Størkersen et al. 2020). State's follow-up of function-based regulation can be an illusion of control—which in the case of DOGE can be seen as through AI technology promising control over how resources are spent, or in *safety management* through verification promising control over how organizations comply with requirements. The term "auditism" symbolizes this, as it points

[1] Example of research references: (Almklov and Antonsen 2014; Andrews et al. 2025; Bieder and Bourrier 2013; Dekker 2025; Hood 2007; Jensen and Winthereik 2017; Lindøe et al. 2011; Power 1994; Rae et al. 2018; Størkersen 2018; Størkersen et al. 2020).

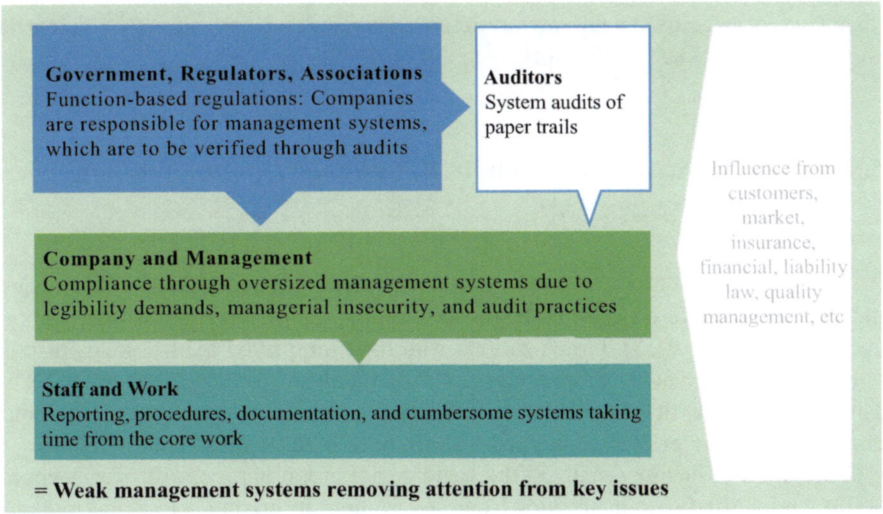

Fig. 5.1 Socio-technical system involved in safety management, illustrating how function-based regulation (deregulation) can produce overregulation with weak management systems that shift attention from key issues. This figure is an original contribution that builds on the ideas illustrated in Rasmussen (1997, p. 1859) and Størkersen and Fyhn (2024, p. 32), but does not reproduce any copyrighted material

at a reaction where organizational actions are targeted at satisfying corporate auditors' demands and result in bullshit tasks and red tape that *hamper* core operations (Rae et al. 2018; Størkersen 2024; Størkersen and Fyhn 2024). This paradox—less state, more red tape—suggests that deregulation does not eliminate regulation but transforms and redistributes it, as Fig. 5.1 illustrates.

5.3 Current and Potentially Future Organizational Responses to Deregulation

Safety management research can serve as a case showing future organizational responses to deregulatory measures. Safety research has over these decades described how function-based regulation results in extensive safety management systems not suiting the operations, but creating counterproductive procedures, exhausting documentation, and "safety work" at the expense of safe work (Bieder and Bourrier 2013; Dekker 2025; Hale and Borys 2013; Rae et al. 2018). Companies handle function-based regulation in a context with existing managerial expectations and cumbersome practices of verification. This also illustrates why current deregulation initiatives, like DOGE and "chainsaw" reforms, are likely to fail in improving business conditions. Deregulation has given rise to new forms of internal bureaucracy and ever-increasing management systems that take attention form real risks and innovation (see our earlier work, such as Dekker 2025; Størkersen 2018).

5.3.1 *Others Step in with* Their *Regulatory Demands When the State Steps Out*

Although function-based regulation states that companies can make simple management systems, more procedures tend to accumulate, particularly after accidents or audits (Amalberti 2013). Organizations who have simplified their systems usually have ended up with at least as large systems after audits (Størkersen and Fyhn 2024). The number of audits has snowballed, as has the documentation and compliance clutter to meet the growing appetite of a variety of other demands and schemes.

Deregulation has externalized the regulatory burden and control over accountability and shifted responsibilities from government to private actors and markets. For example, when DOGE is counting the number of words cut in regulations, it is not necessarily cutting bureaucracy or risks, but moving the focus of regulation to markets and other organizations. Overall, bureaucracy is not reduced, but reshaped and relocated (Dekker 2025). As the state withdraws, private actors step in. For example, if state actors are not giving detailed verification of products, certification companies can provide that service (Almklov et al. 2014). When states reduce regulation of an industry, new organizations and structures are formed (Dean 2025; Nilsen et al. 2022), rarely leading to a reduced administrative burden for the original companies (Ntoumanis et al. 2024).

Legitimate companies are expected by the surroundings to be transparent and accountable (Czarniawska and Sevón 1996). Organizations often mimic each other's compliance strategies, reinforcing bureaucratic norms. A long line of social controls shapes expectations, such as corporate governance, liability law, insurance, and investors (Baram and Lindøe 2013). Specific expectations can come from all types of causes, like universal design, environmental protection, or animal welfare, resulting in new demands and certification schemes (Amundsen 2022). The number of industry standards and management demands is continuingly increasing for each cause getting attention by the international community, just as organizational trends or procedures after accidents (Amalberti 2013; Czarniawska and Sevón 1996). To be sure, management may purchase all-encompassing management systems for each topic, which add to internal bureaucracy (Almklov et al. 2014).

5.3.2 *Companies Complicate Compliance Themselves*

So, we have seen that some deregulation leads to self-induced organizational bureaucratic excess. Although many know that the ill-fitting management systems contradict

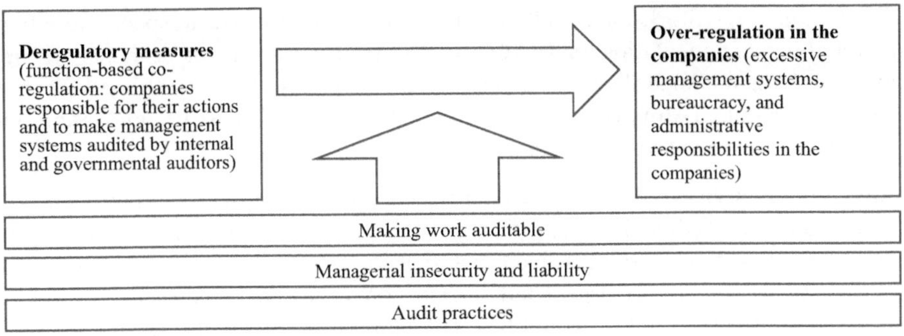

Fig. 5.2 How government deregulation (meant to simplify the administrative burden) is met in organizations and thus leads to overregulation and a private administrative burden. This figure is an original contribution that illustrates the ideas described in Størkersen et al. (2020, p. 202)

or otherwise undermine safety, the pressures to bureaucratize have continued. Below and in earlier work (Størkersen et al. 2020), we suggest the companies' own over-regulation happens because (i) organizations are eager to demonstrate auditability of the variable work; (ii) vague requirements encourage more exhaustive surveillance by insecure management, and (iii) even traditional ways of preparing for audits add to the administrative burden (see also Fig. 5.2).

Making Work Auditable (Legibility): Much function-based regulation requires management systems, with risk assessments and procedures of critical work tasks. Legible work descriptions are key to a management system. In the process of making critical tasks legible—clearly described (Scott 1998), it may be simplified so much the essence is lost. Still, legibility enables auditors to assess compliance through documentation rather than direct observation. Yet, a focus on auditability can distort priorities, encouraging "bullshit tasks" that serve documentation at the expense of safe work (Størkersen 2024).[2]

Managerial Insecurity and Liability: Function-based regulation places heavy responsibility on the organization, without clear criteria for what is sufficient, which again is reinforced by liability law (Power 1994). Function-based rules lack specified compliance thresholds, making organizations vulnerable to liability in the event of failure (Hood 2007). When the regulation state that a company is responsible for safety in its operations, and the most concrete rule is that they must have and maintain a functioning safety management system, it is difficult to know when the company has done enough to prevent accidents (Power 1994; Størkersen et al. 2020). Organizations implement over-engineered systems in fear of legal consequences (Hood 2007), resulting in what practitioners have called a "World Championship in liability

[2] We have, partly seriously, suggested the condition of "auditism" in organizations where the idea of audits shapes how work is structured, performed, or talked about in organizations (Størkersen, 2024; Størkersen et al., 2020). When employees perceive these tasks as meaningless or even "bullshit," yet still perform them, it signals illness.

disclaimers" (see Størkersen et al. 2020). Safety professionals can spend more time maintaining documentation than addressing actual hazards (Provan et al. 2019). They may implement standardized procedures that fail to reflect the complexity of real-world operations (Almklov and Antonsen 2014), as well as stifle innovation and frontline autonomy, creating a false sense of control (Dekker 2025).

Auditors' Practices: Written procedures and documentation are traditionally the only fodder for audits (Hale and Borys 2013, p. 230). Audits are efficient if the management system is standardized, making possible for "activities and situations of each local context being translated into slots on the accountants' sheets" (Almklov et al. 2014, p. 27). If not, it would be challenging to fairly audit a special management system, with procedures particularly made for one single company, based on for example video material or machine logs, in combination with handwritten logbooks. Auditors thus—often without intention—emphasize large, written, standardized management systems that are not perfectly adjusted for the core operations, but for the auditor (Størkersen 2018). For auditors to easily assess internal control systems, it is common that companies design the system specifically to pass audits (Jensen and Winthereik 2017). This means that organizations in practice prioritize and shape safety management systems to fit audits, rather than authentically reflecting operations (Størkersen 2024).

5.3.3 The Potential Contribution of Policymakers (How to Prevent Companies from Suffocating and Deregulation Promoters from Eating Their Hats?)

In today's deregulatory climate—from USA to Argentina to Europe—supporters of leaner regulation may be surprised that such initiatives result in more bureaucracy. This chapter shows that the handling of function-based regulation points at a deregulatory future out of control. To spin the point more positively, the research presented here shows that states have an important role in regulation.

Without holding the political views of leaders like Javier Milei or the current US administration, we have some shared regulatory goals: simplicity, flexibility, innovation, and the removal of ineffective regulations and costs. This will also benefit safety management. But achieving these goals requires a capable and well-resourced government. When companies fail to manage their actual performance and instead produce auditability as a proxy for legitimacy, the involvement of strong governmental actors becomes valuable. Without it, a variety of actors step in to fill the hollowed-out regulatory void—resulting in anything but simplicity and innovation. The barriers to simplification lie in deeply rooted expectations around accountability, verification, and organizational performance, as our research on function-based regulations shows. Deregulatory efforts, which aim to "chainsaw" government functions, also certainly externalize risk without addressing the underlying drivers of organizations.

Simplicity and innovation require improved—not reduced—regulatory resources, as we in this chapter have argued deeply. The façade of accountability maintained by organizations obscures real responsibility and calls for a rethinking of how due diligence is demonstrated. An approach could involve collaboration between private and governmental auditors to develop truer ways of demonstrating "compliance." Auditors could agree not to approve systems leading to internal bureaucracy. "Documentation" could be something that do not demand filing by the personnel.

Through this chapter, we have seen how contemporary deregulatory initiatives create burdens that can choke both companies and regulations' intentions. Policymakers must move beyond the binary of "more" or "less" regulation and focus on how regulatory responsibilities are met in organizations. This may require more—not fewer—resources for regulation, not to expand control, but to ensure regulation gives the intended results.

References

P.G. Almklov, S. Antonsen, Making work invisible: new public management and operational work in critical infrastructure sectors. Public Adm. **92**(2), 477–492 (2014)

P.G. Almklov, R. Rosness, K.V. Størkersen, When safety science meets the practitioners: does safety science contribute to marginalization of practical knowledge? Saf. Sci. **67**, 25–36 (2014)

R. Amalberti, *Navigating Safety: Necessary Compromises and Trade-Offs—Theory and Practice* (Springer, 2013)

V.S. Amundsen, From checking boxes to actual improvement: a new take on sustainability certification. Aquaculture **548**, 737672 (2022). https://doi.org/10.1016/j.aquaculture.2021.737672

D. Andrews, B. Égert, C. Castle, C. De La Maisonneuve, *Regulation and Growth: Lessons From Nearly 50 Years of Product Market Reforms* (OECD Economics Department Working Papers). Organisation for Economic Co-Operation and Development (OECD) (2025). https://doi.org/10.1787/3b3285df-en

R. Baldwin, M. Cave, M. Lodge, *Understanding Regulation: Theory, Strategy, and Practice* (Oxford University Press, 2011)

M.S. Baram, P. Lindøe, Modes of risk regulation for prevention of major industrial accidents, in *Risk Governance of Offshore Oil and Gas Operations*, ed. by P. Lindøe, M. Baram, O. Renn (Cambridge University Press, 2013), pp.34–55

C. Bieder, M. Bourrier, *Trapping Safety into Rules: How Desirable or Avoidable Is Proceduralization?* (Ashgate, 2013)

B. Czarniawska, G. Sevón, *Translating Organizational Change*, vol. 56 (Walter de Gruyter, 1996)

A. Dean, Free trade, repressed workers. New Labor Forum **34**(1), 32–42 (2025). https://doi.org/10.1177/10957960241299780

S. Dekker, *Safety Theater: How the Desire for Perfection Drives Compliance Clutter, Inauthenticity, and Accidents* (Taylor & Francis, 2025)

A.R. Hale, D. Borys, Working to rule, or working safely, in *Trapping Safety into Rules. How Desirable or Avoidable Is Proceduralization*, ed. by C. Bieder, M. Bourrier (Ashgate, 2013), pp. 43–68

C. Hood, What happens when transparency meets blame-avoidance? Public Manag. Rev. **9**(2), 191–210 (2007)

C.B. Jensen, B.R. Winthereik, Audit loops and audit implosion, in *Redescribing Relations: Strathernian Conversations on Ethnography, Knowledge and Politics*. ed. by A. Lebner (Berghahn, 2017), pp.155–181

P. Lindøe, O.A. Engen, A. Moen, Regulatory response to hazards: case studies from the Norwegian petroleum industry, in *Advances in Safety, Reliability and Risk Management*, ed. by A.G. Bérenguer, C.G. Soares (CRC Press, 2011), pp. 1667–1673
M. Nilsen, T. Kongsvik, P.G. Almklov, Splintered structures and workers without a workplace: how should safety science address the fragmentation of organizations? Saf. Sci. **148**, 105644 (2022). https://doi.org/10.1016/j.ssci.2021.105644
A.B. Ntoumanis, S.O. Jorge, E. Zagarese, A. Wunderlin, A.R. Stern, P.J. Luiz, The impact of decree 70/2023 on labor regulation and industrial relations in Argentina: a legal and economic analysis. Int. J. Law Public Policy (IJLAPP) **6**(2), 77–84 (2024). https://doi.org/10.36079/lamintang.ijlapp-0602.694
M. Power, *The Audit Explosion* (Demos, 1994)
D.J. Provan, A.J. Rae, S.W. Dekker, An ethnography of the safety professional's dilemma: safety work or the safety of work? Saf. Sci. **117**, 276–289 (2019)
A.J. Rae, D.J. Provan, D.E. Weber, S.W.A. Dekker, Safety clutter: the accumulation and persistence of 'safety'work that does not contribute to operational safety. Policy Pract. Health Saf. **16**(2), 194–211 (2018)
K.V. Størkersen, T. Thorvaldsen, T. Kongsvik, S. Dekker, How deregulation can become overregulation: an empirical study into the growth of internal bureaucracy when governments take a step back. Saf. Sci. **128**, 104772 (2020). https://doi.org/10.1016/j.ssci.2020.104772
K.V. Størkersen, H. Fyhn, Untangling safety management: from reasonable regulation to bullshit tasks, in *Compliance and Initiative in the Production of Safety: A Systems Perspective on Managing Tensions and Building Complementarity* (Springer, 2024), pp. 29–40
K.V. Størkersen, *Bureaucracy Overload Calling for Audit Implosion: A Sociological Study of How the International Safety Management Code affects Norwegian Coastal Transport* (Norwegian University of Science and Technology, 2018)
K.V. Størkersen, Auditism: symptoms, safety consequences, causes, and cure, in *The Regulator–Regulatee Relationship in High-Hazard Industry Sectors: New Actors and New Viewpoints in a Conservative Landscape* (Springer Nature Switzerland, Cham, 2024), pp. 79–88

Chapter 6
How Can We Make AI Understand Safety?

Malakis Stathis, Baumgartner Marc, Smoker Anthony, and Langejan Erwin

Abstract The integration of artificial intelligence (AI) in safety-critical systems marks a paradigm shift in safety management. Traditional safety frameworks, grounded in deterministic models and compliance-based regulations and rulemaking, face challenges in adapting and evolving to the complexities of integration of AI-driven environments. This chapter examines the epistemic and institutional barriers to AI adoption in safety, emphasizing the role of complexity and tight coupling in system failures. It explores the limitations of current safety management systems (SMS) and AI's challenges, including opaque decision-making and value misalignment. Additionally, it discusses the challenges posed by AI, particularly Large Language Models (LLMs), which introduce risks related to misinformation and interpretability. The chapter argues for a shift toward resilient, adaptive safety frameworks that balance AI autonomy with human oversight. By addressing AI's epistemic constraints and ethical dilemmas, this work proposes strategies for aligning AI decision-making with human safety values.

Keywords Epistemic sovereignty · Work-as-imagined versus Work-as-done · Illusion of safety · Unpredictability · Safety value misalignment

---

M. Stathis
Hellenic Aviation Service Provider, Athens, Greece
e-mail: stathis.malakis@gmail.com

B. Marc (✉)
IFATCA, Geneva, Switzerland
e-mail: marc.baumgartner@ifatca.org

S. Anthony · L. Erwin
Lund University, Lund, Sweden
e-mail: anthonysmoker@icloud.com

L. Erwin
Civil Aviation Authority Norway, Bodø, Norway
e-mail: ela@caa.no

C. Bieder et al. (eds.), *Living with the New Safety Landscape II*,
SpringerBriefs in Safety Management, https://doi.org/10.1007/978-3-032-29343-5_6

6.1 Introduction

Contemporary safety management has been shaped by decades of accumulated knowledge and practice, grounded in conceptions of risk and mitigation strategies that once fitted simpler technological and organizational contexts. These dominant frameworks—rooted in linear causal models, deterministic approaches to hazard analysis, and compliance-based standards—have earned broad acceptance and legitimacy over time. They remain deeply embedded in training programs, institutional policies, regulatory requirements, and professional certifications. This institutional entrenchment, coupled with the socio-political processes that shaped safety thinking, continues to reinforce their enduring presence even as we approach a Kuhnian paradigm shift (Kuhn 1962) precipitated by artificial intelligence (AI) and increasing system complexity.

From an **epistemic standpoint**, these historically dominant safety conceptions owe much to epistemic sovereignty (Rouse 1996). By epistemic sovereignty (Rouse 1996) we mean the institutional authority granted to particular methods and vocabularies of safety. Over decades, regulators, training systems, and certification schemes stabilized linear causal analysis and compliance routines as the legitimate way to know and manage safety, creating path dependence and professional incentives that reinforced their authority. Their dominance emerged and reinforced through familiarity and proven track records that became in essence a "reality of safety" (Healy 2003). Regulatory bodies, industry leaders, and safety practitioners have long relied on these well-documented best practices emanating from sovereign knowledge. This includes linear chain-of-events analyses or step-by-step compliance frameworks—that emphasize clear, causally traceable pathways to failure and prevention and the focus of "standardization" or customized approaches to risk management and governance as well practitioners work through proceduralization and work (Bieder 2023).

In socio-technical systems that were considered "complex" or complicated, this may have analytical authority. These methods are considered to have delivered consistent, empirically validated results: the basis for continuous improvement. Consequently, the community developed a sense of epistemic confidence in these approaches that established a control paradigm of safety (Bieder 2023; Amalberti et al. 2005). They became "settled science," in a Kuhnian sense, whose anomalies were relatively rare and easily explained within the prevailing paradigm.

Complex systems have not been held in stasis but have evolved. Complex adaptive socio-technical systems are a more representative term for the nature of safety-critical systems and critical infrastructures. This therefore challenges the epistemic hegemony safety science as a settled science. The challenge arises from concurrent evolutions: pervasive datafication of operations, machine learning-based decision aids, tighter intersystem connectivity (e.g., networked automation across sectors), higher operational tempo, and climate-driven volatility in traffic and weather. Together, these increase feedback density and shorten decision cycles, stressing assumptions behind linear, document-centered safety practices.

On the societal dimension, there is a strong incentive to maintain stability in safety practices. Because accidents can incur high costs in terms of human life, environmental harm, legal liability, and public trust, drastic shifts in safety philosophy face scrutiny and resistance. Stakeholders, including governments, professional associations, labor unions, and the public expect a certain degree of safety, continuity and reliability. Consequently, the path toward new paradigms is often constrained by institutional inertia and cautious policymaking and epistemic sovereignty as the imperative it brings eschews choice over the universality of knowledge.

Established frameworks offer a predictable set of metrics, roles, and responsibilities for those involved in safety—from frontline practitioners to safety auditors and regulators. Nonetheless, the future legitimacy of these traditional conceptions is being called into question by the unprecedented capabilities and accompanying challenges of AI-driven systems. This evolution undermines the idea that safety knowledge is "settled" and universally applicable and instead demands plural, adaptive ways of knowing and governing safety. What is known as the "illusion of safety" (Bieder 2023; Smith 2018, p. 58). Traditional conceptions of cause-and-effect may prove insufficient for an environment in which machine learning algorithms autonomously optimize processes, generate unexpected emergent behaviors, or adaptively reconfigure system parameters in real time. The challenge to simple cause-and-effect is twofold: (i) learning systems optimize using high-dimensional correlations that are not transparently causal, and (ii) tightly coupled feedback (e.g., demand–capacity balancing interacting with sector configurations and traffic flows) means small algorithmic adjustments can nonlinearly change controller workload and conflict geometry. Illustration (ATM): a ML tool that adaptively suggests minor route or speed advisories to smooth demand can, under atypical weather, concentrate flows and create unanticipated convergence points—risks that are not visible in a linear event chain.

As we move toward integrating cognitive systems engineering principles with advanced analytics and machine learning models, the very nature of risk identification, event prediction, and control strategies changes radically. Such radical change dictates that underlying epistemologies and philosophies change as well. Charles Perrow's ideas (Perrow 1984, 1999) on complexity and tight coupling become increasingly relevant in the era of AI and machine learning as high-speed AI decision-making can propagate errors through system networks with unforeseen speed and scale in virtually any safety-critical system. We use AI governance to mean the structures, roles, and processes that allocate control and accountability for AI behavior across design, deployment, and operation, including assurance cases, monitoring for drift, and decision rights.

This chapter explores the challenges and requirements for AI to function effectively in complex and tightly coupled safety-critical environments (Perrow 1984). We situate AI within wider megatrends—automation, digitalization, climate volatility, and workforce shifts—which collectively pressure traditional safety management. Additionally, we make an initial attempt to examine whether AI should mirror human cognitive models or be allowed to develop safety-related understanding through model-agnostic approaches.

6.2 AI in Safety-Critical Systems: Complexity and Coupling

Safety-critical systems such as aviation exemplify the high complexity and tight coupling described by Normal Accident Theory (Perrow 1984, 1999). Perrow characterised aviation in two discreet forms in Normal Accidents: aircraft being one and 'airways' i.e., air traffic control (Perrow 1999, p. 123). Aircraft is situated in the tightly coupled complex quadrant of the interaction/coupling chart (Perrow 1999, p. 97). Airways however are located close to the center of the chart. "As about as error free a large system as you can find" because "changes could be found that reduced complexity and coupling" (Perrow 1999, p. 168). "Airways" (as system) has intrinsic properties that enable this. While aircraft (as a system) is unlikely ever to eliminate accidents. Airways, it can be argued,as a complex sociotechnical system, can be seen to be challenged by adaptations that eschew order, rationalisation and self organisation and often propagate rapidly due to interdependencies between components. The introduction of AI further increases this complexity by introducing opaque decision-making processes (Malakis et al. 2022).

6.3 Challenges in Teaching AI About Safety

Traditional safety management systems (SMS) have struggled to keep pace with the increasing complexity of modern aviation systems. Especially as noted by Malakis et al. (2023), existing SMS frameworks emphasize compliance and hazard reporting but fail to address emergent risks effectively. AI's integration into SMS must address these shortcomings, particularly the following limitations. Concretely, integrating AI into SMS spans three layers: (1) Capability—analytics/copilots that help detect precursors and synthesize occurrence data; (2) Process—embedding those tools in hazard identification, investigation, and change management (e.g., triaging reports, suggesting bow-ties, surfacing "Work-as-done" patterns); (3) Governance—documented model risk management (assurance cases, versioning, drift monitoring, rollback). The aim is a continuous-learning loop rather than a periodic, document-centered cycle:

- Limited operational attunement. Most AI models lack access to tacit practices and local constraints (e.g., traffic complexity, handover rhythms, phraseology nuances, weather workarounds) that shape *how* safety decisions are made in context. *ATM example:* a model may flag low separation risk on paper yet ignore controller workload spikes from convective weather deviations that degrade safety margins.
- Over-reliance on risk matrices and probabilistic assessments, which fail to capture system complexity (Cox 2008).
- Gaps between "Work-as-imagined" and "Work-as-done," which AI must learn to bridge (Woods and Hollnagel 2006).

- Challenges in handling safety incidents that involve dynamic risk transformations (Kontogiannis and Malakis 2017).

In this line of reasoning the use of AI in air traffic management (ATM) has revealed multiple challenges in ensuring safety (Malakis et al. 2022):

- **Lack of Operational Awareness:** AI lacks an innate understanding of how safety decisions evolve within operational contexts.
- **Challenges in Joint Cognitive Systems:** ATM is a joint cognitive system where human and machine intelligence interact (Hollnagel and Woods 2005). The integration of AI requires seamless communication and adaptability between air traffic controllers and AI-driven decision aids.
- **Opaque Decision-Making:** AI often operates as a "black box," making it difficult for safety practitioners to understand how it derives conclusions.

6.3.1 Value Alignment and the Safety Paradigm Shift

One of the critical challenges in AI safety is the **value alignment problem** (Russell 2019). AI must align its decision-making processes with human safety values, but defining these values in computational terms is challenging. This is particularly challenging in the specific context of ethical and moral dimensions. Attempts to define human purpose may imbue machine artifacts with objectives or goals, Russell argues, "that are imperfectly aligned with the human" (Russell 2019, pp. 137–138).

Safety is context-dependent and requires calibrated adaptation with traceable guarantees. While modern AI can vary outputs with prompts or data, this is not the same as governable adaptation that preserves safety constraints across shifting operational contexts. In practice, more context-sensitive adaptation increases the burden on control and assurance: The system must both flex to local context and remain auditable, testable, and bounded by human-set constraints. In addition to the value alignment, recent work emphasizes the importance of control–accountability alignment among AI stakeholders for mitigating AI risks (Grote et al. 2024). According to this perspective, control enables effective influence over AI outcomes, while accountability motivates stakeholders to employ that control in line with societal and organizational values (Grote et al. 2024). However, AI systems' adaptive capabilities, including self-learning from large datasets, can reduce predictability and traceability of behavior over time (as models update or encounter drift), even when design-time transparency is acceptable. As well as constraining and lengthening accountability in real-time choices and strategy formulation.

To address these gaps, Grote et al. (2024) propose a decentralized form of governance, bringing together AI developers, users, and senior management so that each stakeholder can negotiate and align control with accountability. Perrow (1984, 1999) recognized the need for decentralized structures in organizations working with high-risk technologies. But controversially advocated a need for structures that were held centralized and decentralized agency in authority, stakeholders are encouraged to engage in integrative rather than distributive negotiations, aided by perspective taking and extensive information sharing. By institutionalizing accountability as a shared norm, powerful actors—such as AI providers who own large data and computing resources—are more likely to relinquish some control or accept greater accountability, thereby reducing risks of unchecked AI behavior.

In this context, the following key issues must be addressed:

- **Interpretability:** AI models must provide explanations for their safety-related decisions.
- **Ethical Considerations:** AI must make trade-offs in safety-critical scenarios, necessitating a robust ethical framework. Ensuring that LLMs align with established safety principles and ethical frameworks remains a challenge (Gabriel 2020).
- **Human Oversight:** AI should enhance rather than replace human judgment in safety-critical environments.
- **Control–Accountability Alignment:** A shared sense of accountability among all parties (developers, users, and management) must be matched by giving each the necessary control to mitigate risk effectively (Grote et al. 2024).

6.3.2 Model-Agnostic Versus Human-Like AI Learning

A key research question is whether AI should learn safety in a way that mirrors human cognition or develop its own model-agnostic approach. A model-agnostic AI may identify new safety insights beyond human intuition, but it risks developing strategies that are difficult for humans to interpret. Conversely, a human-like approach ensures that AI's safety reasoning aligns with existing safety paradigms but may limit AI's ability to discover novel safety enhancements. It is imperative that balancing these two approaches is essential for effective AI integration in safety-critical domains.

6.3.3 Safety Assurance and AI Certification

One of the major challenges in AI-driven safety systems is **certification and regulatory assurance**. Unlike traditional aviation systems, where certification follows a structured and well-defined process, AI-driven safety solutions present novel challenges. According to the European Aviation Safety Agency (EASA 2023), certifying AI-based systems requires new regulatory approaches, including:

- **Transparency and Explainability:** AI decisions must be interpretable and justifiable within safety regulations.
- **Robustness and Resilience:** AI must demonstrate reliability in high-stakes environments without unpredictable behavior.

- **Human–AI Collaboration:** Certification processes must ensure that AI complements rather than replaces human oversight.

EASA (2023) has highlighted that the aviation industry must develop new standards and frameworks to evaluate AI safety and performance before deployment in safety-critical operations and they work diligently in this direction. This challenge underscores the need for an evolving certification methodology that accounts for the dynamic nature of AI learning and decision-making processes.

6.4 Discussion and Conclusion

Large Language Models (LLMs) and other advanced AI systems show potential to assist human experts with complex cognitive tasks (e.g., synthesizing evidence for risk assessment, drafting investigation narratives), rather than to automate high-stakes decision authority. However, their application in safety-critical domains is accompanied by several limitations:

- **Lack of Real-World Understanding:** LLMs generate responses based on statistical patterns rather than genuine comprehension of safety contexts (Bender and Koller 2020).
- **Hallucination and Misinformation:** LLMs may fabricate plausible but incorrect information, which poses risks in high-stakes environments (Bommasani et al. 2021).
- **Interpretability Challenges:** The reasoning process behind LLM-generated outputs is difficult to trace, complicating accountability in safety-critical operations (Marcus 2020).

While LLMs have the potential to enhance safety processes, their deployment in real-time or near real-time decision-making roles within aviation must be approached cautiously. Robust validation mechanisms, human-in-the-loop oversight, and regulatory scrutiny are essential to mitigate the risks associated with their use in safety-critical systems.

The increasing role of AI in safety-critical systems necessitates a re-evaluation of how AI can be trained to understand safety. This paper provides one perspective of AI Governance, that recognizes that there is potential, perhaps transformative, for governance and management of safety to evolve utilizing the capacity that digital transformation and AI-based systems offer. It is arguably inevitable that the nature of such human-system integration will be co-agency rather than indeterminate hybrids; this is "Governance with AI," which is distinctly different from the "Governance of AI: the governance of designs of complex adaptive systems that exploit AI such as human-autonomy systems in digital transformation".

This paper draws on Grote et al.'s (2024) propositions in the assertions argued in this chapter. Which can be seen as the basis of a paradigmatic change in the meaning of safety itself and the time-honored custom and practice of safety management.

Given the limitations of traditional SMS and the complexity of modern systems, AI must be designed to adapt to emergent risks, align with human values, and operate transparently. Building on recent theories of control–accountability alignment (Grote et al. 2024), safety-oriented AI design may involve decentralizing governance and engaging multiple stakeholders—from frontline operators to senior management—in integrative negotiations that emphasize perspective taking and iterative feedback loops.

Another promising approach to achieving a more integrated and human-centric AI ecosystem is explored by Mollick (2024), who introduces the concept of "co-intelligence." In this framework, human workers and AI systems collaborate as symbiotic partners, each bringing complementary strengths to problem-solving and decision-making. Rather than conceiving of AI merely as a tool to replace or augment human labor, co-intelligence emphasizes dynamic collaboration. Humans contribute situational awareness, ethical judgment, and domain expertise, while AI provides high-speed computational capacity and pattern recognition.

By treating AI as an active "co-creator" of safety solutions, the co-intelligence model adds a constructive layer to the strategies discussed in earlier sections. For instance, it can reinforce control–accountability alignment by clearly delineating human domains of decision-making (e.g., setting ethical boundaries) while letting AI handle the data-intensive aspects of safety analytics.

According to Mollick (2024), organizations that adopt co-intelligence typically:

- **Encourage open channels of communication** so that AI-driven insights can be questioned or enhanced by human operators.
- **Develop feedback loops** where both the AI system and the human team learn from real-world data, feeding into continual improvements in safety processes.
- **Invest in training programs** that foster "AI literacy," ensuring that human experts understand not just the outputs of AI but also, to some extent, the models' logic and limitations.

In safety-critical environments, the co-intelligence approach may serve as an additional safeguard: By pairing human values with machine intelligence in real time, emergent anomalies could be identified more quickly, and decisions made more robustly. This human–AI tandem has the potential to reduce the black-box effect and promote a shared sense of accountability, ultimately leading to safer operational practices.

In this stride, recent research by Cummings (2025) underscores how AI can introduce new points of failure when deployed in safety-critical applications, particularly in scenarios where traditional accident models do not fully account for opaque or adaptive algorithms. Cummings proposes the Taxonomy for AI Hazard Analysis (TAIHA), which adds four AI-specific "layers" that can lead to accidents or near-misses:

- **Inadequate AI Oversight Factors** such as high commercial pressure or insufficient regulatory policies can incentivize firms to deploy AI systems prematurely, sidestepping essential safety protocols.

- **Inadequate AI Design Errors** in how data are fused, or how neural networks are trained, may leave the system incapable of recognizing unusual but foreseeable conditions.
- **Inadequate AI Maintenance** because AI models can degrade when environmental conditions shift (model drift), they must be continuously updated and audited.
- **Inadequate AI Testing,** AI-based systems in a single, controlled setting or relying solely on simulations can mask deeper performance issues that only arise under real-world complexity.

Cummings (2025) argues that meaningful human oversight must focus on thorough testing and transparent certification processes—akin to the role of a chief test engineer in aerospace—so that each AI deployment is deemed "safe enough" to operate. In practice, this means clarifying where accountability resides along these layers, so that if an AI model fails in an unexpected way, there is a clear mechanism to trace back which stakeholders—developers, testers, senior managers—are responsible for addressing and correcting the failure.

Together with the co-intelligence framework (Mollick 2024) and control–accountability alignment (Grote et al. 2024), TAIHA highlights the importance of proactively managing complex, adaptive AI systems. These theoretical advances converge on a single message: Emergent hazards require emergent forms of governance, wherein AI's unique risks are acknowledged and integrated into every stage of design, oversight, and continuous maintenance.

In conclusion, the future of AI in safety requires a paradigm shift that goes beyond mere compliance-based models toward dynamic, resilient, and interpretable AI-driven safety management. By bringing together insights from cognitive systems engineering, robust governance structures, and the evolving landscape of certification requirements, we can create AI systems that not only comply with procedural knowledge but also align with human and societal values. Such alignment will be vital to ensure that AI remains a trusted ally in tackling the increasing complexity and coupling present in today's safety-critical domains.

References

R. Amalberti, Y. Auroy, D. Berwick, P. Barach, Five system barriers to achieving ultrasafe health care. Ann. Intern. Med. **142**(9), 756–764 (2005). https://doi.org/10.7326/0003-4819-142-9-200505030-00012

E.M. Bender, A. Koller, Climbing towards NLU: on meaning, form, and understanding in the age of data, in *Proceedings of the 58th Annual Meeting of the Association for Computational Linguistics* (2020), pp. 5185–5198. https://doi.org/10.18653/v1/2020.acl-main.463

C. Bieder, *Safety Management Systems and their Origins* (CRC Press, 2023)

R. Bommasani et al., On the opportunities and risks of foundation models (2021). arXiv:2108.07258. https://doi.org/10.48550/arXiv.2108.07258

L.A. Cox, What's wrong with risk matrices? Risk Anal. **28**(2), 497–512 (2008). https://doi.org/10.1111/j.1539-6924.2008.01030.x

M.L. Cummings, *Identifying AI Hazards and Responsibility Gaps* (IEEE Access, 2025).
European Union Aviation Safety Agency (EASA), *Artificial Intelligence Roadmap 2.0* (European Union Aviation Safety Agency, 2023). https://www.easa.europa.eu
I. Gabriel, Artificial intelligence, values, and alignment. Mind. Mach. **30**, 411–437 (2020). https://doi.org/10.1007/s11023-020-09539-2
G. Grote, S.K. Parker, K. Crowston, Taming artificial intelligence: a theory of control–accountability alignment among AI developers and users. Acad. Manage. Rev. (2024)
S. Healy, Epistemological pluralism and the politics of choice. Futures **35**, 689–671 (2003)
E. Hollnagel, D.D. Woods, *Joint Cognitive Systems: Foundations of Cognitive Systems Engineering* (Taylor & Francis, 2005)
T. Kontogiannis, S. Malakis, *Cognitive Engineering and Safety Organization in Air Traffic Management* (Taylor & Francis, 2017)
T.S. Kuhn, *The Structure of Scientific Revolutions* (University of Chicago Press, 1962)
S. Malakis, T. Kontogiannis, A. Smoker, A pragmatic approach to the limitations of safety management systems in aviation. Saf. Sci. (2023) https://doi.org/10.1016/j.ssci.2023.106215
S. Malakis, M. Baumgartner, N. Berzina, T. Laursen, A. Smoker, A. Poti, G. Fabris, Challenges from the introduction of artificial intelligence in the European air traffic management system. IFAC-PapersOnLine **55**(29), 1–6 (2022). https://doi.org/10.1016/j.ifacol.2022.09.440
G. Marcus, The next decade in AI: four steps towards robust artificial intelligence (2020). arXiv: 2002.06177. https://doi.org/10.48550/arXiv.2002.06177
E. Mollick, *Co-Intelligence: Living and Working with AI* (Random House Penguin, 2024)
C. Perrow, *Normal Accidents: Living with High-Risk Technologies* (Princeton University Press, 1984)
C. Perrow, *Normal Accidents: Living with High-Risk Technologies—Updated Edition* (REV-Revised) (Princeton University Press, 1999)
J. Rouse, *Beyond Epistemic Sovereignty* (Division I Faculty Publications. Paper 37, 1996). http://wesscholar.wesleyan.edu/div1facpubs/37
S. Russell, *Human Compatible: Artificial Intelligence and the Problem of Control* (Viking, 2019)
G. Smith, *Paper Safe: The Triumph of Bureaucracy in Safety Management* (2018)
D.D. Woods, E. Hollnagel, *Joint Cognitive Systems: Patterns in Cognitive Systems Engineering* (CRC Press, Boca Raton, FL, 2006)

Chapter 7
Supporting Excellence in Professional Practice in a Changing Industrial Landscape

Jan Hayes and Sarah Maslen

Abstract Frameworks for engineering competence focus on technical skills, whereas non-technical skills are also key for excellence in professional practice for safety. Our framework for non-technical capabilities, developed for expert pipeline engineers in Australia, identifies six capability groups: use foresighted reasoning; understand norms and values; think systematically; collaborate with others; ground decisions in reality; and advocate for action. The increased pace of industrial change introduces significant epistemic risks for three worker groups, including the existing professionals moving to work with new technology, new professionals joining the sector, and existing professionals left to manage legacy assets. Each faces their own epistemic challenges, but the framework of non-technical capabilities provides a basis for addressing them and ensuring ongoing safe performance.

Keywords Engineering competence · Non-technical capabilities · Engineering practice · Safety decision-making · Epistemic risk

7.1 Introduction

The changing industrial landscape calls for professionals with the skills to navigate rapid change in both technology and working arrangements while still making the best safety choices. In addition to a high level of technical competence, safety decision-makers in complex hazardous industries need strong capabilities in a range of non-technical areas to make the best choices. Recent research has focused on identifying a comprehensive set of such capabilities in the context of pipeline engineering (Hayes et al. 2021; Holdsworth et al. 2022). Many engineers value these skills and see them as informing their professional judgment. Despite this, they are

J. Hayes (✉) · S. Maslen
RMIT University, Melbourne, Australia
e-mail: jan.hayes2@rmit.edu.au

S. Maslen
e-mail: sarah.maslen@rmit.edu.au

C. Bieder et al. (eds.), *Living with the New Safety Landscape II*,
SpringerBriefs in Safety Management, https://doi.org/10.1007/978-3-032-29343-5_7

rarely discussed as many organizations still believe safety is a matter of "reliable equipment, good procedures and processes, well-behaved operators" (Bieder and Bourrier 2013, p. 3). In the following, we briefly set out past safety research on non-technical skills and capabilities. We then describe our research on non-technical capabilities before addressing how these capabilities support engineering judgments in the changing industrial landscape.

7.2 Beyond Technical Competence

The foundations of robust decision-making have been studied within different domains of practice. In aviation, crew resource management (CRM) programs have promoted training for non-technical skills in small teams operating in an aircraft cockpit (Flin et al. 2008) and air traffic management in a control tower (Read and Charles 2018) with a focus on small group communication and decision-making in time-pressured environments. Many safety-critical decisions are made in situations that do not conform to the temporal conditions of CRM cases, with the impact of actions not felt for many years. Reason's Swiss cheese model (1997) posits that the unsafe acts of individuals are caused by workplace factors such as the physical environment and shift patterns which are themselves determined by organizational factors such as budgets and organizational structures. This framing underplays the agency of professionals, although Reason later proposed that organizational accident sequences can be broken by what he called "error wisdom" on the part of individual workers (2004). He had suggestions of some skills as part of this error wisdom, but did not develop a comprehensive framework. Other scholars offer further clues in their analysis of what supports the necessary chronic unease on the part of senior managers: propensity to worry, pessimism and the cognitive abilities of requisite imagination, flexible thinking and vigilance (Fruhen and Flin 2016).

While each of these bodies of academic work attempts to describe capabilities for safe outcomes, they have not taken engineering technical specialists as their focus. In contrast, various professional societies have developed capabilities or competency frameworks themselves. For instance, the Institution of Chemical Engineers Safety Centre have published the IChemE Safety Centre (ISC) Process Safety Competency Guidance (The Institution of Chemical Engineers 2018) which sets out 18 topics. Process safety competency is also an element of the US Center for Chemical Process Safety's (CCPS) risk-based process safety management system (Center for Chemical Process Safety 2007). These frameworks focus strongly on technical competence, although the broader literature tells us this is but part of the picture.

7.3 Capabilities for Decision-Making

In light of this gap in understanding, our research further developed what capabilities professional pipeline engineers need in order to perform their role effectively while navigating a complex organizational, as much as technical, context involving many other actors including managers, clients, contractors, field personnel, peers, and other external stakeholders including government regulators, housing developers, and community stakeholders. Their work focuses primarily on design and integrity management—activities with long-term, rather than immediate, impacts. In addition to technical know-how, achieving excellent outcomes requires non-technical skills that we term "capabilities." This notion of capabilities is deliberately broader than pure technical competence, incorporating excellence in professional practice (Stephenson 1998). Attending to individual professionals in this way does not downplay the importance of organizational factors in accidents, viewing individual capability as an essential component of organizational safety as a whole.

The capabilities shown in Table 7.1 were developed through interviews with 41 pipeline engineers working in the natural gas sector in Australia (for a complete discussion of the research and the framework see Hayes et al. 2021).

The first capability group recognizes that decisions made now can have safety implications long into the future. Professionals need to understand the consequences of their actions and imagine multiple futures. This means recognizing the importance of understanding the worst potential consequences when making decisions (Capability 1). Decision choices should be made with those consequences in mind (Capability 2). In addition, pipeline engineers often make decisions in conditions of uncertainty, especially when trying to consider the longer term (Capability 3).

Professional practice for pipeline engineers is influenced by three norms, and these are addressed in the second capability group: the standard that applies to their technical work (Capability 4), company and legislated requirements for consideration of risk (Capability 5), and professional ethics (Capability 6).

The third capability group recognizes that the work of any individual engineer is part of a bigger whole. The first capability about thinking systematically is a need to understand the context of decision-making (Capability 7). The second is the need for pipeline engineers to know how to prioritize actions in complex situations (Capability 8). Once the broader context of an issue is understood, pipeline engineers often draw on risk processes and concepts as a way of describing interconnectedness and developing priorities (Capability 9). A final aspect of thinking systematically is being thorough and taking time to make the right choice (Capability 10).

Group D capabilities link to the collective nature of engineering work addressing relationships with other people in one's own profession and beyond. Engineering work is rarely a solitary exercise, with engineers consulting others including members of their own engineering discipline, and other stakeholders (Capability 11). Related to this is knowing one's professional limitations and when to call on others (Capability 12). A key aspect of this capability group is building good relationships with field personnel (Capability 13). Many pipeline engineers have roles that also benefit from

Table 7.1 Pipeline engineering capabilities for safety

Capability group	Capability from interview
A. Use long-term, foresighted reasoning, especially in the face of uncertainty	1. Imagining worst case outcomes 2. Choosing with the long term in mind 3. Developing strategies to deal with uncertainty
B. Understand norms and values that inform actions	4. Interpreting standards beyond black letter compliance 5. Understanding risk processes and concepts 6. Making ethical choices
C. Think systematically and understand interconnectedness	7. Understanding operational, business, legal, and cultural contexts 8. Prioritizing actions in complex situations 9. Applying risk concepts 10. Taking enough time
D. Collaborate with and draw on the experience of others	11. Building networks or seeking others' opinions 12. Knowing your limitations 13. Building a good relationship with field personnel 14. Building external stakeholder relationships
E. Ground decisions in reality	15. Making practical choices 16. Learning from small failures 17. Being skeptical of models and simulations 18. Imagining being there
F. Advocate for action and take responsibility	19. Influencing senior management and clients 20. Standing up for public safety

building relationships with external stakeholders including councils, other utilities, specialist suppliers, developers, and regulators (Capability 14).

Group E capabilities are about linking the theoretical and the practical, i.e., the need to make decisions that are practical (Capability 15), understanding what has gone wrong in the past (Capability 16), and being skeptical of the outputs of models and simulations (Capability 17). It can also involve an imaginative check (Capability 18).

The final group of capabilities relates to taking action. In many cases, pipeline engineers need to convince those more senior in their organization of the need to act on matters that can impact public safety since more senior people are in control of the necessary resources (Capability 19). Conversely, taking action to try to stop something that has a potential adverse impact on safety outcomes is also important (Capability 20).

These capabilities have been used as the basis for a major experiential, case-based learning program delivered in the Australian pipeline sector in the period 2022–2024. The program was delivered with a research focus on evaluation of its effectiveness (Holdsworth et al. 2022). Over 150 pipeline engineers took part. Evaluation activities three months after the training showed that participants typically could not name the non-technical capabilities, but in their discussion of the cases in interview, they showed a knowledge of them, i.e., the generic lesson is bound to the narrative, rather than recalled abstractly (Maslen et al. 2025).

7.4 Changing Industrial Landscape

Over the next decade, workers in hazardous industries will be faced with an increasing pace of change. In the context of climate change, some industries are declining, some are expanding rapidly, and others are changing in fundamental ways as they decarbonize their activities. Other technology innovations such as AI are accelerating changes already underway due to increased digitalization. Globalized supply chains are also impacting industries as they compete in a global market. These strategic trends will increasingly impact workers' employment arrangements, and ultimately, the skills required from workers to do their work successfully.

This changing landscape introduces a range of new socio-technical risks to public safety that can arise from structural, organizational, technological, epistemic, or cultural sources (Macrae 2022). In the pipeline context, we see that many risk sources are epistemic issues that have implications for the decision-making of individual professionals, given the lack of clarity and even contestation over the technical requirements of hydrogen blends, limited expertise in working with the material across the industry, and significant time pressures placed on the transition. Under these conditions, the capability framework for non-technical skills becomes even more important. We observe three different groups of workers (existing professionals moving to work with new technology, new professionals joining the sector and existing professionals left to manage legacy assets), who each face their own epistemic challenges.

As the gas pipeline companies move to decarbonize, existing professional staff face major knowledge challenges with their professional practice. Collectively, staff have only limited experience in designing and operating systems with future fuels and some of the basic scientific data (particularly the properties of blended fuels) are still unknown. This raises critical questions about what is known and what is not. Theoretical scientific knowledge about the physical properties, behavior, and so management of pure hydrogen and hydrogen/methane blends is a critical foundation to engineering practice. In the context of the energy transition, engineers need to keep up to date with knowledge contained in standards (Capability 4) and be aware of the limits to their expertise (Capability 12), particularly in the context of some significant uncertainties in basic data (Capability 3).

We also know that field experience is vital to making sound decisions and so a lack of field experience is another epistemic risk. Mitigation of this risk relies on maximizing the exposure to design and operating experience as new facilities come online. This includes activities such as secondments, networking, setting up advisory arrangements, and similar things that give engineers as much exposure to as many new facilities as possible to build relevant experience and bring it back to be used in design and operation of local facilities. Formal systems at an industry or organizational level are important here but so is the attitude of individual professional engineers. Capabilities in Groups D and E play a strong role here in recognizing the value of both field experience and collaboration with colleagues.

In the context of limited practical experience with new technologies, the inevitable small faults and failures that accompany any practical work become particularly important for learning. Such small anomalies, faults, and errors embody a "gift of failure" (Carroll and Fahlbruch 2011). Rather than being fixed and then forgotten, each fault is a potential source of valuable insights into system health (Capability 16). Small faults and failures not only provide valuable lessons about system behaviors. They can have an important social function in giving regular reminders of the potential for major disaster and every individual's role in preventing such. In the excitement that goes with new technology, early success can be taken as a sign that all is well. In the worst case, those involved lose the ability to imagine worst case scenarios and so the potential for disaster (Pidgeon and O'Leary 2000). Group A capabilities encourage use of long term, foresighted reasoning and counter this tendency to be overenthusiastic.

The nature of capital project work always emphasizes the short-term goal of on time, on budget completion but the energy transition is aiming to proceed at an even more accelerated pace than usual. Placing such a premium on meeting deadlines means that, in project organizations, speed is celebrated as "a synonym of good" (Czarniawska 2013, p. 11). Capabilities in Group C counter this tendency by encouraging engineers to think systematically and understand interconnectedness, including the need to take enough time for decision-making in complex situations.

Communicating safety messages to senior management (Capability Group F) also remains important, particularly in an environment where CEOs and Board members are increasingly professionals with functional backgrounds in finance, accounting, management, and law, rather than technical professionals who have been promoted up through the company (Phillips and Cotter 2010).

We see that current pipeline sector engineers who have been trained in the 20 non-technical capabilities will have a solid foundation on which to draw as the technical basis of their professional practice shifts. The sheer volume of work created by the energy transition will necessarily involve many new people moving into the sector who may or may not have these capabilities. Making cultural and competency assumptions about new cohorts of people could be a significant source of risk in the energy transition. This will need to be mitigated by training in non-technical capabilities along with explicit communication of expectations regarding safety attitudes and performance. This must be reinforced by monitoring and feedback of actions to ensure that any early deviations from expectations are identified and addressed.

Conversely, the creation of new business areas within the gas pipeline companies will necessarily be done in parallel with continuing to operate gas facilities that are coming to the end of their operational life. There is potential for risk associated with these assets to increase as organizational focus moves elsewhere. This is partly a structural issue, but people working on stranded assets can feel left out and anxious as they face a professional existential threat of their job disappearing and their expertise being no longer valued. The study by Song et al. (2024) of the impact on metro train drivers of moves to AI and driverless trains is a direct illustration of the negative impact that workplace anxiety can have on job performance. This could contribute directly to increased failures of existing assets when management and professional attention is focused elsewhere. Risks in this area are primarily mitigated by the actions of senior management in ensuring that those working on legacy assets feel that they are valued by the organization, i.e., that their work is significant to a successful energy transition, but the non-technical capabilities have a role here too by emphasizing professional ethics (Capability 6) and retaining a focus on worst case outcomes (Capability 1).

In summary, non-technical capabilities are recognized as a key part of what goes to make an expert pipeline engineer in Australia. While the capabilities were developed based on current engineering practice, consideration of epistemic risks linked to the future of the industry shows their continuing importance for the future of safety in the Australian pipeline industry.

Ethics Statement This study was performed in line with the principles of the Declaration of Helsinki. Approval was granted by the RMIT University College Human Ethics Advisory Network (project number 2020-23458-10862 dated 28 July 2020). Informed consent to participate in the research and for data to be published was obtained from individual participants.

References

C. Bieder, M. Bourrier, (eds.) *Trapping Safety into Rules: How Desirable or Avoidable is Proceduralization?* (Ashgate, 2013)

J.S. Carroll, B. Fahlbruch, "The gift of failure: new approaches to analyzing and learning from events and near-misses". Honoring the contributions of Bernhard Wilpert. Saf. Sci. **49**, 1–4 (2011)

Center for Chemical Process Safety, *Guidelines for Risk-Based Process Safety* (Wiley Interscience, 2007)

B. Czarniawska, Is speed good? Scand. J. Manag. **29**, 7–12 (2013)

R. Flin, P. O'Connor, M. Crichton, *Safety at the Sharp End: A Guide to Non Technical Skills* (Ashgate, 2008)

L.S. Fruhen, R. Flin, 'Chronic unease' for safety in senior managers: an interview study of its components, behaviours and consequences. J. Risk Res. **19**(5), 645–663 (2016)

J. Hayes, S. Maslen, S. Holdsworth, O. Sandri, Defining the capable engineer: non-technical skills that support safe decisions in uncertain, dynamic situations. Saf. Sci. **141** (2021). https://doi.org/10.1016/j.ssci.2021.105324

S. Holdsworth, J. Hayes, O. Sandri, S. Maslen, Developing professional expertise for safety: a learning design framework. Cogn. Technol. Work **24**, 459–472 (2022). https://doi.org/10.1007/s10111-022-00702-6

C. Macrae, Learning from the failure of autonomous and intelligent systems: accidents, safety, and sociotechnical sources of risk. Risk Anal. **42**(9), 1999–2025 (2022). https://doi.org/10.1111/risa.13850

S. Maslen, J. Hayes, O. Sandri, S. Holdsworth, "What stuck with me was...": How sources and approaches to case based learning affect participant recall three months later. Saf. Sci. **191**, 106952 (2025). https://doi.org/10.1016/j.ssci.2025.106952

P.J. Phillips, J. Cotter, The technostructure gap: the educational qualifications of executive and non-executive directors. Corp. Ownersh. Control. **7**(4A), 102–113 (2010)

N. Pidgeon, M. O'Leary, Man-made disasters: why technology and organizations (sometimes) fail. Saf. Sci. **34**, 15–30 (2000)

K. Read, R. Charles, Understanding teamwork errors in royal air force air traffic control. Saf. Sci. **109**, 36–45 (2018). https://doi.org/10.1016/j.ssci.2018.04.030

J. Reason, *Managing the risks of organizational accidents* (Ashgate, 1997)

J. Reason, Beyond the organisational accident: the need for "error wisdom" on the frontline. Qual. Saf. Health Care **13**, 28–33 (2004). https://doi.org/10.1136/qshc.2003.009548

K. Song, M. Guo, L. Ye, Y. Liu, S. Liu, Driverless metros are coming, but what about the drivers? A study on AI-related anxiety and safety performance. Saf. Sci. **175**, 106487 (2024). https://doi.org/10.1016/j.ssci.2024.106487

J. Stephenson, The concept of capability and its importance in higher education, in *Capability and Quality in Higher Education*, 2nd edn, ed. by J. Stephen, M. Yorke (Routledge, 1998)

The Institution of Chemical Engineers, *ISC Process Safety Competency Guide (Edition 2, 2018)* (2018). https://www.icheme.org/media/12452/0007_18-competency_brochure-update.pdf

Part II
Learning from Cases and Examples

Chapter 8
Health Care in Crisis: Prospects and Directions

Rene Amalberti, Siri Wiig, and Charles Vincent

Abstract The healthcare sector entered a major crisis 10 years ago, at the forefront of other safety-critical industries. This chapter first summarizes the crisis of the combined impact of an aging population, growing demand, shrinking workforce and their combined impact patient safety. The chapter then examines the various responses which have evolved and which continue to evolve. The health sector is already confronting the problem of "living with" enormous pressures. This is what makes this sector so interesting for this book, because we can already draw several lessons beyond health, for industrial safety.

Keywords Health care · System change · Chronic staff shortage · Complexity · Rapid pace of innovation · Solution for adapting

8.1 Introduction

Health care is a complex service by its very nature, because of the number and professions of carers involved, the continuity of care across many settings, the natural variety of patients and their pathologies, and the difficulty of providing compassionate care in an often highly pressured environment. Healthcare systems also vary enormously across the world in terms of their level of resource, organization, and sources of funding. In addition, the extremely rapid pace of innovation, both for technology, therapeutics, and organizational change, has no equivalent in other industries. Innovative therapies are applied to an ever-growing, more fragile, older population

R. Amalberti (✉)
Foncsi, France
e-mail: rene.amalberti@foncsi.org

S. Wiig
University of Oxford, Oxford, UK
e-mail: siri.wiig@uis.no

C. Vincent
University of Stavanger, Stavanger, Norway
e-mail: charles.vincent@psy.ox.ac.uk

C. Bieder et al. (eds.), *Living with the New Safety Landscape II*,
SpringerBriefs in Safety Management, https://doi.org/10.1007/978-3-032-29343-5_8

with growing care needs and a consequently higher risk of medical error over long periods of time.

8.2 The Risks of Health Care

There is a huge literature converging internationally since the famous IOM report "To err is Human" published in 1999, showing that the risk of harm in hospital is considerable. Studies in many countries reviewing medical records have shown that about 10% of patients admitted to hospital suffer some harm, defined as requiring at least one extra day in hospital. About half of these events are judged to be preventable with current care. The harm may be minor and quickly resolved, but some patients suffer very severe adverse events (de Vries).

The most recent studies of adverse events in the USA of adverse events have found little change in the overall rates over time, finding that 7% of hospitalized patients experienced a preventable adverse event in the calendar year 2018 (Bates et al. 2023). However, studies using electronic medical records present a rather different picture. For instance, Eldridge (2022) used the Medicaid EMR to review twenty-one specific harms 2012–2019 for AEs in relation to AMI, heart failure, pneumonia, and major surgery. They found that an overall reduction in harm (drug events, infections, falls, surgical events) though gave little detail on which events reduced most. A number of other studies have also shown a fall in serious adverse events suggesting that rates of some serious adverse events may be reducing, even though overall rates of harm may be stable or even increasing due to new risks being introduced. Most studies of adverse events have however been hospital based, and we have much less understanding of levels of harm in primary and community care and almost no information on the safety of care in the home.

Safety in health care, unlike other industries, is difficult to track reliably partly because it is a "moving target" in the sense that the expected standard changes over time (Vincent and Amalberti 2015). The overall picture suggests that the overall rate of events in hospitals at least may be increasing or at least not reducing, but the events that do occur are either less serious or detected and mitigated more quickly.

8.3 Health Care Under Pressure: The Current Crisis

Health systems are under more pressure than ever before, and the challenges are multiplying and accelerating. Economic forces, new technology, genomics, AI in medicine, increasing demands for care are all playing a part or are predicted to increasingly do so. Above all, aging populations in many parts of the world are exacerbating the pressures on the system and intensifying the requirements to provide effective care equitably to citizens.

The greatest challenge facing health care is a shortage of skilled workers, across all professions particularly in deprived areas. Healthcare systems have always faced shortages of skilled staff at times, but these problems have markedly increased in the post COVID era even in well-resourced systems. This is partly because demand for health care has outstripped capacity, but also because the pressures on staff are growing (Kirk 2024). Working conditions are deemed to be outdated and burdensome, which has led to a steady loss of motivation and increased chance of early retirement or change of profession. The aspirations of workers have also changed, with greater demands for quality of life and work/life balance, which has made health care less attractive as a career. There is little hope of a return to previous staffing levels, nor of limiting the growth in demand for care linked to the massive aging of the population.

For patients and families, the most immediate signs of pressure are the difficulty of accessing health care, the near collapse of primary care in some countries, lack of coordination, and fragmentation of care. Everything is happening as if COVID—outside the COVID period itself—was followed by a period of brutal acceleration in the deterioration of care, especially primary care, specifically a collapse of what can be called "compassionate medicine": listening, being available, providing local and quality care in hospital and at home, simply finding a professional (access). The greatest risk to patients now may be the difficulty of accessing care at all, rather than errors made while receiving care.

8.4 How Can Health Care Respond to the Current Crisis?

Health care has drawn inspiration from the traditional industrial model to improve safety. The safety of "high-technology specialized medicine" and healthcare technologies are approached as complex socio-technical systems akin to industrial installation. The experience of patients however, which may not be closely correlated to the risks from high-technology medicine, has more in common with industrial quality and customer experience. The experience of healthcare professionals in contrast, particularly psychological safety, has more in common with quality of working life in industrial settings.

Over the last 20 years, many of the tools used in industrial safety have been borrowed and applied to the medical field to reduce risk using a variety of levers to control the healthcare safety and its accelerating costs. These include defining best practices, continuous quality improvement, reporting of undesirable events, training in teamwork, and safety culture improvement among others. There is wide-ranging evidence that these approaches can produce improvements to the safety and quality of care in specific settings, though less evidence of widespread dissemination and impact (Refs). However, the optimistic vision of earlier years and steady improvement of safety and quality is threatened by the increasing pressures on health care which both increase risk to patients directly and disrupt improvement and innovation (Amalberti and Vincent 2020). The healthcare sector has now to focus more on managing risk than on improving safety, in the sense of learning to live with high demand and inadequate resources to provide the level of care that is expected.

The rest of the text presents various "living with" solutions that have evolved in response to the crisis facing health care in recent years. These approaches vary in how explicitly they are formulated, how much they form part of an official narrative, and the extent to which they have been tested on a greater or lesser scale. These solutions are presented in three main sections. The first expresses solutions that are compatible with the fundamentals of industrial safety today, the idea that safety results from defense against expected and foreseeable risks, conduct governed by the rules of good practice (in normal and abnormal situations) all within a dedicated organization that learns from its mistakes (safety management system). The second set accepts the basic premise of a defined system working in largely predictable ways but recognizes that adaptations of various degrees will have to be made to manage risk and maintain productivity. The third section sets out solutions that challenge the current safety model, going beyond it and reconsidering its fundamentals. The last section, and to some extent the second, describes practices that are evolving in response to pressures but are mostly not officially sanctioned by either by organizations or by regulators.

8.5 Solutions Compatible with the Fundamentals of Industrial Safety Today

Health care, like other industries, would ideally like to expand the numbers of skilled staff to meet the demands of the aging population. This however will take time and, even if achieved, there will at the very least be a period where current standards cannot be met. If standards are still to be met, then one approach is to adapt the standards to the new reality.

8.5.1 Expand the Base of Professionals to Resolve the Understaffing Crisis

All countries are trying, in the longer term, to expand the healthcare workforce to meet demand. The primary approach is to increase numbers in training within the home country. Developed nations have been reliant on recruiting staff from poorer countries for many years, and immigration has sustained the British NHS and many other systems. This approach is not sustainable in the long-term for a variety of reasons, not least because it deprives emerging countries of their clinical workforce.

Many countries have also taken another approach. They have started to broaden the range of professions and actors authorized to visit and prescribe, massively opening it up to less qualified jobs, hence maintaining and redeploying the core technical medical profession to essential tasks. This has already started with the advent of professions that were not directly involved in the consultations, i.e., physician assistant (trained 5 years instead of 7–9), certified nurses, or clinical pharmacists, all authorized to visit and prescribe under supervision of physicians. But the biggest expansion could come from care being provided, in the home in particular, by informal carers (family or friends) or unskilled paid carers with no clinical training.

Several states are ahead and have widely started to deploy these solutions, particularly the UK, the USA, and Canada. For example, delivery of primary healthcare services by nurses instead of doctors probably leads to similar or better patient health and higher patient satisfaction. Nurses probably also have longer consultations with patients. Using nurses instead of doctors makes little or no difference in the numbers of prescriptions and tests ordered (Cochrane Review 2018). The impact on long-term health outcomes and cost is less clear (Abrams et al. 2020). Similar findings have been reported from Canada, Australia, the USA, and other countries. The feedback almost all converges toward two results: concern from the professionals replaced about quality of care and funding shortages, but satisfaction from patients who consider that these solutions at least provide access to some health care even if they are not able to see a doctor (Greenhalgh and McKee 2025).

8.5.2 *Enforce Compliance with Rules While Adapting Quality Targets to What Is Realistically Achievable*

The logic behind this solution is essentially to lower expectations of the healthcare system, while continuing to enforce protocols and compliance with rules. This means continuing to use audits and recommendations, quality controls, and quality improvement, but redefining the standards to become attainable targets rather than adhering to a "one size fits all" maximum requirements. For instance, a recent Australian initiative estimated achievable benchmarks of care (ABC) for twelve quality indicators used to monitor long-term care services (LTCS) and to identify characteristics associated with attaining the estimated ABC. Smaller and government LTCS were generally more likely to attain the realistic goals of ABC (Ervin et al. 2023).

Standards can be adjusted to a modest degree but, to cope with the pressures on health care, this adjustment needs to be combined with a willingness and ability to maintain and improve standards while under pressure. This approach relies on five key capacities which need to be fostered within organizations and systems. First, an ongoing effort to build and deepen capacity and capability for improvement. Second, careful nurturing and development of a culture and context that provides staff and service users with autonomy, skills, and permission to assess creative ideas. Third, leadership behaviors that create psychological safety and reinforce problem-solving at the point of care, supported by policy initiatives and incentives for high quality. Fourth, a robust infrastructure to ensure that teams have close, skilled support as they tackle some of the most complex challenges in the organization (Shah and Sunol 2024). Finally, leaders of systems and regulatory organizations need to accept and support this modified vision which combines adjustment of standards to a realistic level with continuing long-term improvement.

8.6 Adaptations to Manage Risk and Maintain Productivity

Adaptations take many forms in health care and occur at all levels of the system, from direct patient care, management of the system, and regulatory approaches. Here we illustrate this general approach with two specific examples.

8.6.1 Adaptations in Clinical Work

This approach accepts that adaptations and non-compliance to "classic rules" will be the norm to deliver safe care in different contexts (Clark et al. 2024). Safety cannot be achieved only by following rules, especially in a system under pressure. Rather, the task is to communicate and coordinate adaptations and guide operators toward the most acceptable and safest workarounds (which are becoming new norms).

This path, inspired by and extensively described in the resilience literature, can only be effective if three important conditions are met: (1) organizations are able to implement and handle both positive and negative feedback about the impact of adaptations, (2) teams and organizations need to evaluate and assess the value of deviations having let to good outcomes, (3) healthcare systems need to accept and develop training for professionals in these winning strategies for workarounds, while still maintaining essential safety standards.

This solution challenges a dogma of classic safety by accepting that we will not return to a normal situation of staffing and working conditions and that we must accept individual and collective adaptation efforts to follow safe and effective rules, but which are inevitably more demanding in terms of working conditions (Clark et al. 2024). This solution is already widely practiced in the field, generally without risk to patients and even with potential benefits, but is only rarely officially supported by organizations and regulators (Page et al. 2024). However, when a patient is harmed, staff and organizations will be more vulnerable and open to criticism precisely because they have deviated from accepted standards of care, even if that seemed in the patient's best interests at the time.

8.6.2 Regulatory Resilience: Increasing the Capacity for Safe Adaptation

Effective regulation is, in many industries, one of the foundations of safety. However, time- and resource-consuming regulatory compliance can compromise the ability of staff to be flexible and can potentially be detrimental to safety if it adds to pressure on staff. Compliance measures can thus be anti-resilient (Wiig and O'Hara 2021). However, regulation can potentially shift from being compliance-based to recognizing the need for adaptation and potentially designing and developing the regulatory regime in ways that may contribute to increasing the capacities of adaptation, anticipation, and learning (Furnes Øyri et al. 2024). This approach is complementary to the clinical adaptations discussed in the previous section; regulators to accept the reality of adaptations and to consider how they may be harnessed and managed to maintain both safety and productivity.

The overall message here is that regulators, inspectors, healthcare organizations, managers, healthcare professionals, patients, and families are sources of resilience, implying that the regulators need to offer incentives for reflexivity and collaborative involvement in regulatory design and implementation. Bringing people together in reflexive spaces has great potential to foster reflections about current challenges and adaptations in daily work practices. Regulatory resilience bets that regulators may contribute to creating better conditions for resilient performance in health care if they adapt that model rather than rely on a prescriptive model (Wiig et al. 2023).

8.7 More Radical Solutions

Other directions for health care can be envisaged, some relying on innovation and others which require a reshaping of the fundamentals of health care. We illustrate possible approaches by considering a more radical approach to resilience and the possibilities offered by artificial intelligence.

8.7.1 Radical Resilience Founded on Individual Autonomy

The general literature on contingency management (HROs, resilience engineering, complexity theories) advocates training operators to manage contingencies more effectively by adopting specific organizations within the group of workers and meta-strategies (such as the five tenets of HROs) and/or by giving staff greater autonomy to deviate from rules and practices which are impeding the delivery of care or which may even be threatening safety.

The classic high-reliability organizations are founded on basic procedures and standards, but with freedom to deviate when circumstances demand. This thinking could potentially be extended to give much greater individual autonomy to all healthcare staff. This would imply relying much more extensively on preparing and training healthcare professionals to manage safety more fluidly and dynamically by, for instance, adopting the tenets of high-reliability organizations (Weick and Sutcliffe 2001) and adaptative teamwork strategies. The hope would be that the risks of the inevitably departures from standards would be more than compensated for by the increased safety awareness and error recovery skills of a more autonomous workforce.

This approach is much more radical than that described above, because it would give authority and autonomy to individuals to break rules as often as they consider useful rather than adapt within a previously authorized framework. The HRO model relies on a very strong foundation of standard procedures and practices and a willingness to return to the standard model when pressures ease. Health care arguably does not have this foundation, being already poorly prescribed and organized in many areas, so further autonomy might threaten the system still further. This autonomous approach would move away from the prescriptive nature of safety at an unpredictable rate, is more radically opposed to the present standards, and would open a whole new field of legal and liabilities issues. In addition, patients and families might well lose trust and confidence in a system with increasingly varying and unpredictable forms of care.

8.7.2 Change the Offer by Massively Adopting Generative AI Technology

The most immediate benefits of AI might be seen in providing more efficient access to care and guidance to patients and staff. AI-based platforms could quickly take over effectively for advice—orientation—reassurance and guiding patients in primary care for ad hoc appointments. In a way, this would provide 24/7 advice and triage, potentially significantly reducing emerging risks while massively saving workforce (Howell 2024).

A network of existing high-tech connected surveillance tools (cameras and sensors) could also be extensively deployed at home to improve a continuous monitoring of risks (cardiac, complications of all nature, falls, drug compliance, etc.) (Gallo et al. 2024). New AI generative tools are still experimental in health care and show significant limitations in some clinical applications (Goodman et al. 2024). However, the results of these experiments are converging and already highly positive, showing an even better satisfaction of patients.

Proponents of AI suggest that this class of AI-based systems is ready for implementation and simply awaits official sanction from authorities and arrangements for funding and payment. AI could reduce the administrative burden on healthcare staff and so release time for direct clinical care and potentially to take on many complex clinical tasks, but may well also introduce new forms of risk. However, the overall impact of AI on patient care and cost-effectiveness will not become clear for many years.

8.8 Conclusion

Health and safety have had to adapt even more quickly than other industries to changing societal paradigms. In this sense, health care is already pointing the way to directions and improvements that deserve to be considered in other sectors. Some approaches, such as relying more on unskilled workers, have already been extensively trialed and to some extent evaluated. Some other approaches, such as stepping back from prescriptive safety rules and allowing greater autonomy, are not formally recognized or evaluated. Approaches of this latter kind have evolved rather than been planned, and both the positive and detrimental effects are largely hidden from view. A further risk which needs to be monitored is that solutions may favor technical transformation over human transformation, with the paradoxical risk of falling into systems that are increasingly efficient and safe, but which are poor at providing care and compassion.

What remains to be done is to define each approach more precisely and then consider what combination of all these possibilities might provide both a feasible and coherent set of solutions without an excessive trade-off between productivity and safety. Most importantly, the success or failure of any approach, especially those with a high reliance on technology, ultimately needs to be judged on whether patients experience care and compassion. In this particular regard, health care of course remains unique.

References

R. Abrams, G. Wong, K.R. Mahtani, S. Tierney, A.M. Boylan, N. Roberts, S. Park, Delegating home visits in general practice: a realist review on the impact on GP workload and patient care. Br. J. Gen. Pract. **70**(695), e412–e420 (2020)

R. Amalberti, C. Vincent, Managing risk in hazardous conditions: improvisation is not enough. BMJ Qual. Saf. **29**, 60–63 (2020)

D.W. Bates, D.M. Levine, H. Salmasian, A. Syrowatka, D.M. Shahian, S. Lipsitz, J.P. Zebrowski, L.C. Myers, M.S. Logan, C.G. Roy, E. Mort, The safety of inpatient health care. N. Engl. J. Med. **388**(2), 142–153 (2023)

D. Clark, R. Lawton, R. Baxter, L. Sheard, J.K. O'Hara, Do healthcare professionals work around safety standards, and should we be worried? A scoping review. BMJ Qual. Saf. (2024). https://doi.org/10.1136/bmjqs-2024-017546

Cochrane Review, Nurses as substitutes for doctors in primary care. Cochrane Database Syst. Rev. (7). Art. No. CD001271 (2018). https://doi.org/10.1002/14651858.cd001271.pub3

N. Eldridge, Y. Wang, M. Metersky, S. Eckenrode, J. Mathew, N. Sonnenfeld, J. Perdue-Poli, D. Hunt, J. Brady, P. McGann, E. Grace, D. Rodrick, E. Drye, H.M. Krumholz, Trends in adverse event rates in hospitalized patients, 2010–2019. Jama **328**(2), 173–183 (2022)

J. Ervin, Y. Taouk, B. Hewitt, T. King, The association between unpaid labour and mental health in working-age adults in Australia from 2002 to 2020: a longitudinal population-based cohort study. Lancet Public Health **8**(4), e276–e285 (2023)

S. Furnes Øyri, J. Braithwaite, D. Greenfield, S. Wiig, Resilience and regulation–antithesis or a smart combination for future healthcare service improvement? Int. J. Qual. Health Care **36**(1), mzae002 (2024). https://doi.org/10.1093/intqhc/mzae002

R.J. Gallo, L. Shieh, M. Smith et al., Effectiveness of an artificial intelligence-enabled intervention for detecting clinical deterioration. JAMA Intern. Med. **184**(5), 557–562 (2024). https://doi.org/10.1001/jamainternmed.2024.0084

K.E. Goodman, H.Y. Paul, D.J. Morgan, AI-generated clinical summaries require more than accuracy. JAMA **331**(8), 637–638 (2024). https://doi.org/10.1001/jama.2024.0555

T. Greenhalgh, M. McKee, Physician associates and anaesthetic associates in UK: rapid systematic review of recent UK based research. BMJ **388** (2025)

M.D. Howell, Generative artificial intelligence, patient safety and healthcare quality: a review. BMJ Qual. Saf. (2024). https://doi.org/10.1136/bmjqs-2023-016690

K. Kirk, Time for a rebalance: psychological and emotional well-being in the healthcare workforce as the foundation for patient safety. BMJ Qual. Saf. 1–4 (2024). https://doi.org/10.1136/bmjqs-2024-017236

B. Page, D. Irving, R. Amalberti, C. Vincent, Health services under pressure: a scoping review and development of a taxonomy of adaptive strategies. BMJ Qual. Saf. **33**(11), 738–747 (2024)

A. Shah, R. Sunol, What does the future of quality improvement look like? Int. J. Qual. Health Care **36**(3), mzae070 (2024)

C. Vincent, R. Amalberti, Safety in healthcare is a moving target. BMJ Qual. Saf. **24**, 539–540 (2015)

K.E. Weick, K.M. Sutcliffe, *Managing the Unexpected*, vol. 9 (Jossey-Bass, San Francisco, 2001)

S. Wiig, H.B. Lyng, V. Guise, E. Ree, B. Fagerdal, H. Dombestein, L. Schibevaag, J. Braithwaite, C. Haraldseid-Driftland, From theory to policy in resilient health care: policy recommendations and lessons learnt from the resilience in health care research program. J. Patient Saf. **20**(7), e109-e114 (2023)

S. Wiig, J.K. O'Hara, Resilient and responsive healthcare services and systems: challenges and opportunities in a changing world. BMC Health Serv. Res. **21**, 1–5 (2021)

Chapter 9
Lost in Sustainability? What Happens to Road Safety When Integrated in the Broader Sustainability Agenda?

Tor-Olav Nævestad and Hanna Wennberg

Abstract Since the adoption of the UN Sustainable Development Goals (SDGs) in 2015, road safety has been reframed as part of sustainable development. The 2020 Stockholm Declaration, launching the UN's Second Decade of Action for Road Safety (2021–2030), emphasizes road safety as a prerequisite for other SDGs. This chapter examines whether integrating road safety into broader sustainability frameworks is beneficial. SDG integration allows road safety to align with more politically prioritized policy areas, such as public health, workplace safety, and environmental and climate policies. This enables cross-sector alliances, new policy arenas, and access to funding. At the same time, road safety risks may be diluted when combined with competing sustainability goals. We conclude that road safety is more likely to be lost without SDG integration, as sustainability transitions in society and transport systems will proceed regardless. Reframing road safety within sustainability contexts may therefore strengthen its role as a policy concept.

Keywords Road safety · Sustainability development goals · Vision Zero · Safe System

---

T.-O. Nævestad (✉)
Institute of Transport Economics, Oslo, Norway
e-mail: Tor-Olav.Naevestad@toi.no

H. Wennberg
Lund University, Lund, Sweden
e-mail: hanna.wennberg@tft.lth.se

C. Bieder et al. (eds.), *Living with the New Safety Landscape II*,
SpringerBriefs in Safety Management, https://doi.org/10.1007/978-3-032-29343-5_9

9.1 Introduction

The United Nations Sustainable Development Goals (SDGs)[1] constitute a global framework for sustainable development, adopted by the UN General Assembly in 2015. The SDGs serve as a common platform and set of objectives for countries, businesses, and civil society to eliminate poverty, combat inequality, and handle climate change and other environmental issues. The SDGs are integrated and indivisible and balance the three dimensions of sustainable development: the economic, social, and environmental.

Following the establishment of the SDGs, road safety has been reframed as an integrated part of, and precondition for, other SDGs. Through the 2020 Stockholm Declaration[2] and the related UN Resolution 74/299, road safety, specifically mentioned in SDG 3.6 and 11.2, is seen as integral to achieving the broader goals of the 2030 Agenda for Sustainable Development. The 2020 Stockholm Declaration marked the start of the UNs second global decade of road safety for 2021–2030. This not only aims to reduce deaths and injuries by 50% by 2030, but it also redefines road safety as a key enabler for addressing other societal challenges and achieving other objectives, for example, related to public health, climate, and social inclusion and equity. Thus, the Stockholm Declaration underlines the role of road safety as a precondition for other sustainability ambitions, e.g., positive health outcomes by active mobility (related to SDG 3), more attractive cities (related to SDG 11), climate, local emissions, and energy-use (related to SDGS 13, 7, and 15), reduced inequalities (related to SDG 10), and gender equality (related to SDG 5). The expert recommendations for the 2020 Stockholm Declaration (STA 2020) refer to this understanding of road safety as a "sustainability development goal integration approach" focusing on the positive synergies between road safety and other SDGs. We refer to this as the "Integrated approach."

The importance of leadership for road safety by public and private actors is emphasized in the Stockholm Declaration. Organizations' responsibilities are further expressed by the expert recommendations for the 4th Global Ministerial Conference on Road Safety in Marrakech in 2025. While the 2020 expert recommendations focus on integration of road safety in the sustainability agenda and point to road safety as part in sustainability practices and reporting and public procurement (STA 2020), the 2025 expert recommendations revolve around occupational road safety, especially in workplace safety regulations and practices (STA 2025).

The aim of this chapter is to discuss whether and how it is beneficial (or not) for road safety as a policy concept to be a part of the broader sustainability contexts. We discuss two alternatives: (1) SDG integration allows for road safety to "hitch the wagon" to other policy areas and implementation contexts, often more politically prioritized, and (2) Road safety may be at risk of being "lost in sustainability," when combined with other considerations. Before discussing this, we describe existing

[1] https://sdgs.un.org/goals.

[2] https://bransch.trafikverket.se/en/startpage/operations/Operations-road/vision-zero-academy/global-ministerial-conferences-on-road-safety/.

discourses of road safety: the ("old") Vision Zero and Safe System discourse and the ("new") Sustainability agenda discourse (Integrated approach).

9.2 Discourses of Road Safety Policy

9.2.1 *Road Safety in the Vision Zero and Safe System Discourse*

Road safety implementation is in Sweden and Norway carried out systematically in the direction of Vision Zero, based on an established model for management-by-objectives. This approach is in the international literature referred to as a Safe System approach in which road safety management is the foundation involving multisectoral partnerships, lead agencies, and systematic management by PDCA (plan, do, check, act) process of continuous improvement (Nævestad et al. 2025; ITF 2022). Thus, Vision Zero also involves a Safe System approach referring to a system perspective in implementation and road safety management.

Vision Zero, adopted in Sweden in 1997 and in Norway in 2001, involved a new understanding of system designers' responsibility in addition to a road user responsibility (i.e., shared responsibility), a new understanding of a road system adapted to the human body's tolerance level and cognitive limitations (i.e., forgiving road transport system), and a focus on the prevention of fatalities and serious injuries in road traffic (Kristianssen et al. 2018).

Previous studies show that adaptation of Vision Zero led to substantial improvements in road safety from 2000 to 2020. Norway experienced a 73% reduction in traffic fatalities from 2000 to 2020 while Sweden experienced a 63% reduction (Nævestad et al. 2025). These countries have the fewest traffic fatalities per million inhabitants in the world (Belin 2021). The reduction in traffic fatalities plateaued in countries leading in road safety during 2010–2020. As the success of Vision Zero may make road safety appear less urgent compared to other societal challenges, Reitan (2023) argues in favor of focusing on the contribution of road safety to other purposes.

9.2.2 *Road Safety in the Sustainability Agenda Discourse*

Ma et al. (2021) present linkages between injury prevention in general and other SDGs based on a literature review concluding that "the impact is bi-directional, as achievement of the SDGs cannot occur without reducing injury, while reducing injury will assist in realising the SDGs" (Ibid, p. 170). Four main opportunities for injury prevention by realizing synergies with other SDGs are discussed: by (1) land-use planning and urban design, (2) age-friendly cities preventing falls and child-friendly streets initiatives, (3) safeguarding water systems in low- and middle-income countries, and (4) corporate social responsibility in companies and organizations.

These opportunities for road safety integration in the SDG agenda are also largely captured by the 2020 Stockholm Declaration. The 2020 expert recommendations (STA 2020) recommend a target of 50% reduction in road deaths and serious injuries by 2030, based on expanded application of the five Safe System pillars, adoption of Safe System principles, and integration of road safety among other SDGs. A set of nine recommendations were proposed to contribute to further reductions in road deaths and injuries:

1. Sustainable Practices and Reporting: Including road safety interventions across sectors as part of SDG contributions.
2. Procurement: Utilizing the buying power of public and private organizations across their value chains.
3. Modal Shift: Moving from personal motor vehicles toward safer and more active forms of mobility.
4. Child and Youth Health: Encouraging active mobility by building safer roads and walkways.
5. Infrastructure: Realizing the value of Safe System design as quickly as possible.
6. Zero Speeding: Protecting road users from crash forces beyond the limits of human injury tolerance.
7. 30 km/H: Mandating a 30 km/h speed limit in urban areas to prevent serious injuries and deaths to vulnerable road users when human errors occur.
8. Technology: Bringing the benefits of safer vehicles and infrastructure to low- and middle-income countries.
9. Safe Vehicles Across the Globe: Adopting a minimum set of safety standards for motor vehicles.

At least the first two recommendations concern the roles of companies and organizations in relation to occupational road safety. The organizational context is characterized by many positive relationships between road safety and other sustainability issues from an organizational perspective, such as climate and energy efficiency, safe work environment, image and customer satisfaction, and economy (Wennberg and Hyllenius 2024; Nævestad et al. 2023). Several of the recommendations concern the roles of transport authorities and urban planning departments for which the positive synergies between SDGs are central. For example, lower speeds zones and mandating 30 km/h speed limit in urban areas is positively correlated with reducing accidents and injuries, emissions and noise levels, and traffic congestion (Yannis and Michelaraki 2024).

9.3 SDG Integration as an Opportunity to Hitch the Wagon

There are several synergies between road safety as a policy concept and other SDGs, as identified by Wennberg and Dahlholm (2024) in a study of road safety implementation by local authorities in Sweden (Fig. 9.1).

These policy synergies indicate synergies at the outcome level, indicating, e.g., that integrated policies will have positive outcomes for several SDGs simultaneously (e.g., emissions, accidents, gender equality). This justifies integrated policies, aiming to fulfill several SDGs simultaneously.

This indicates that SDG integration allows for road safety to "hitch the wagon" to other policy areas and implementation contexts, often more politically prioritized. For example, public health goals can justify 30 km/h zones. Climate mitigation can go together with eco-driving initiatives involving speed compliance. Road safety becomes a part of workplace health and safety (SDG 8.8) and is embedded in sustainability practices and reporting (SDG 12.6) and public procurement (SDG 12.7). Road safety communication may also be a part of mobility management strategies. SDG integration of road safety provides chances to build new alliances across sectors and

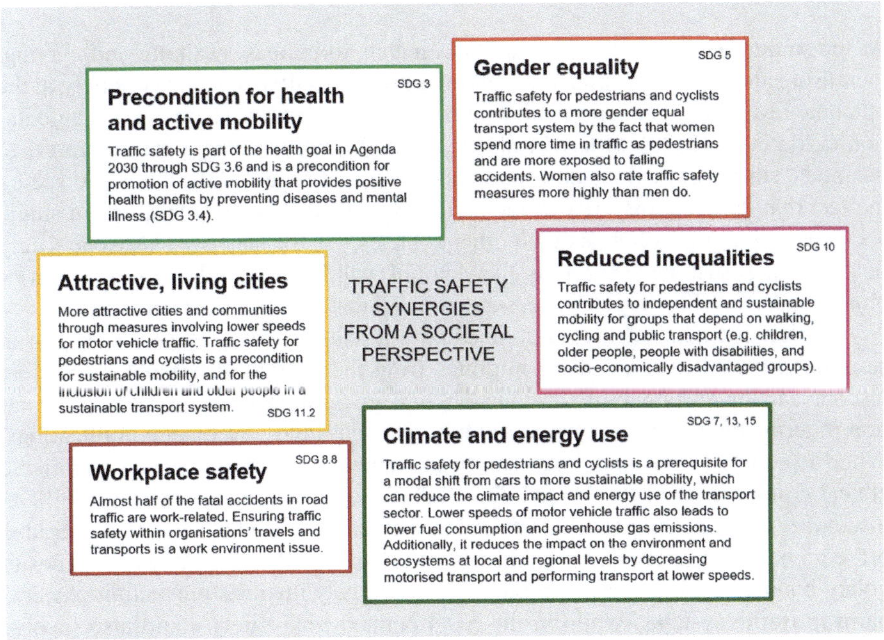

Fig. 9.1 Overview of synergies between road safety and other SDGs; adapted from "Overview of synergies between traffic safety and other sustainability goals (SDGs)" originally published in Wennberg, H. and Dahlholm, O. Traffic safety implementation at local level in line with 2030 Agenda for Sustainable Development: synergies, challenges and enablers. Traffic Safety Research 5, https://doi.org/10.55329/lgfl1703 (2024); released under a Creative Commons Attribution 4.0 International License (CC BY 4.0)

opens new policy doors when road safety goes hand in hand with other concerns and ambitions. This may facilitate implementation of measures positive to road safety, provide access to new funding sources and policy channels.

Even though road safety is only explicitly recognized in SDG 3.6 and SDG 11.2, the relevance of the SDGs for improving road safety is much broader. By reframing road safety within broader global priorities, an advancement toward SDG achievement can be accomplished, while simultaneously reducing fatalities and seriously injured in road traffic.

From an injury prevention perspective, the SDG agenda offers "an unprecedented opportunity to integrate safety considerations across all sectors and stakeholders" (Ma et al. 2021, p. 166). However, integration will not take place in practice if not all stakeholders are aware of road safety being a part of sustainability and how to manage road safety within their sphere of influence (Wennberg and Hyllenius 2024; Wennberg and Dahlholm 2024). This communicative task is in the interest of road safety advocates to address.

9.4 Can Road Safety Be Lost in Sustainability?

At the same time as road safety can be a means to increase walking and cycling, increasing amounts of cycling and walking imposes challenges for road safety at the outcome level (i.e., more single bicycle accidents and pedestrian falls). Argumentation in Ma et al. (2021) with reference to the mobility literature, generally refers to the modal shift from car to cycling as an opportunity for road safety, but is overseeing the fact that the great majority of the seriously injured cyclists are injured in single accidents and not in collisions with other vehicles. On the negative side, road safety outcomes may also be affected by the share of walking and cycling. These injuries must be addressed by targeted measures beyond the mobility shift strategies.

Additionally, goal integrated policies pose a risk that road safety can be lost in sustainability. When road safety migrates from the Vision Zero and Safe System discourse to the Sustainability discourse, it may go through a conceptual transformation in terms of framing, scope, normative priorities, and governance implications. When it comes to framing, road safety is framed as an absolute non-negotiable ethical commitment in the Vision Zero/Safe System discourse, while in the SDG discourse, safety becomes one of several priorities that must be negotiated or traded off, e.g., between safety and modal shift, or urban densification. When it comes to policy boundaries, road safety in Safe System largely involves preventing physical harm in traffic systems, while in the SDG context road safety could also involve including mobility justice (Who gets to be safe? Are vulnerable groups protected?) and environmental safety (e.g., reducing harm not just to humans, but to ecosystems). When it comes to indicators and accountability, road safety in the Safe System context involves, e.g., a focus on the number of deaths and serious injuries, fatality rates, time-bound goals (e.g., halve deaths by 2030), and accountability is direct and outcome-focused. In the SDG context, the metrics may include more holistic indicators, e.g., perceived safety, equity of safety outcomes, safety as part of Quality-of-Life indicators. Thus, road safety becomes one variable in larger sustainability dashboards and could potentially be downplayed in favor of other outcomes, unless protected as a core target.

Finally, there are also important governance differences. Although road safety management in the Safe System context involves many stakeholders, it may be easier to define accountability for one outcome than several outcomes (SDG goals), involving many lead agencies. In the sustainability context, the leadership will likely be even more scattered among various actors in public and private sectors. It is necessary to "trust the processes" in other implementation contexts, but it is in the interest of road safety advocates to gain an overview of the actors and processes in relevant implementation contexts. For some contexts, there is also a need for guidance and tools on how to include and handle road safety together with other SDGs—as is currently being undertaken through the FIA Road Safety Index for companies' sustainability practices and reporting (FIA 2023). Hence, a central lead agency for road safety with responsibility to coordinate and facilitate action, innovation, and collaborations among actors is a necessity.

9.5 Conclusions: Road Safety May Be Lost Without SDG Integration

Is road safety at risk of being "lost in sustainability"? On the one hand, sustainability integration imposes potentials for achieving road safety targets along with other SDGs, while on the other hand, this also requires road safety to be considered in implementation contexts beyond the traditional ones. In fact, road safety may be "lost without SDG integration" since the sustainability transition of our society and transport systems will go on anyway. The sustainability transition of society and transport systems is necessary for many reasons and will go one whether road safety is onboard or not. The inclusion of road safety in the broader sustainability context is therefore a matter for survival for road safety. In other words, road safety can rather be "found" by being reframed into the sustainability context.

However, to avoid road safety being lost in sustainability, several policy challenges need to be addressed successfully. The first is related to integration of several goals at once, dealing with conflicts and/or synergies between outcomes. This requires mechanisms to deal with conflicts and synergies. Conflicting interests are inevitable, and must be managed throughout planning, and this management concerns processes and collaborations between policy areas. Wennberg and Dahlholm (2024) presents key enablers for a sustainability integration approach for road safety, for example, to use platforms shared with other target areas and base implementation on indicators and measures shared with the other areas. In local authorities, such a key platform for road safety implementation as part of sustainability is the context of Sustainable Urban Mobility Planning (SUMP). In companies, sustainability practices and reporting are an example of such platform. However, the sustainability context is currently underutilized due to low awareness among companies about their impact on road safety and the fact that road safety is a sustainability issue (Wennberg and Hyllenius 2024).

The second challenge is related to keeping road safety as an absolute value, to avoid road safety being downplayed in favor of other concerns, or in other words: to avoid road safety being lost in sustainability. The integration of road safety in the broader sustainability agenda may change narratives, implementation contexts, and prioritizations, but it does not change road safety as such. The approach and direction of Vision Zero remain intact, and the integration is more a matter of a necessity to keep advancing road safety implementation in society and to achieve road safety targets together with other societal goals and ambitions. Even though road safety is treated as an integrated part of sustainability, and as a precondition for other SDGs, road safety has still a value of its own: to save lives and reduce negative health loss related to road accidents.

The third challenge is related to communicating the role of road safety as part of sustainability. There is a great potential for reducing fatalities and seriously injured in road traffic by means of sustainability integration by leveraging the positive (bi-directional) synergies with other SDGs. However, this potential is underutilized due to lack of awareness of road safety being a sustainability issue and a lack of joint processes and collaborations between policy areas.

References

M.Å. Belin, The Swedish Vision Zero—an advanced safety culture phenomenon, in *Transport and Safety. Springer Transactions in Civil and Environmental Engineering*, ed. by G. Tiwari, D. Mohan (Springer, Singapore, 2021). https://doi.org/10.1007/978-981-16-1115-5_1

FIA, FIA Road Safety Index—Framework. Federation Internationale de l'Automobile: Paris, France (2023). https://www.fia.com/fia-road-safety-index

ITF, The Safe System Approach in Action. International Transport Forum (2022). https://www.itf-oecd.org/sites/default/files/docs/safe-system-in-action.pdf

A.-C. Kristianssen, R. Andersson, M.-Å. Belin, P. Nilsen, Swedish Vision Zero policies for safety - A comparative policy content analysis. Saf. Sci. **103**, 260–269 (2018). https://doi.org/10.1016/j.ssci.2017.11.005

T. Ma et al., Out of the silos: embedding injury prevention into the sustainable development goals. Inj. Prev. **27**(2), 166–171 (2021). https://doi.org/10.1136/injuryprev-2020-043850

T.-O. Nævestad, J. Blom, V. Milch Uhlving, Sikkerhet og sirkularitet: Sammenheng mellom økonomi, trafikksikkerhet, kundetilfredshet, klima og miljø i busstransport. TØI-rapport 1981/2023. Institute of Transport Economics: Oslo, Norway (2023). https://www.toi.no/publikasjoner/sikkerhet-og-sirkularitet-sammenheng-mellom-okonomi-trafikksikkerhet-kundetilfredshet-klima-og-miljo-i-busstransport

T.-O. Nævestad, E.F. Sam, H. Farah, D. Mwamba, J. Masaki, A. Laureshyn, M. Magnusson, A. Varhelyi, R. Elvik, J. Blom, L.E. Egner, T. Miyoba, L.S. Bisht, Safe system implementation in three African and three European countries: preliminary results from a comparison of six countries. Transp Res Procedia **89**, 243–254 (2025)

T. Reitan, Att styra mot noll: Förverkligandet av Nollvisionen i trafiken. Södertörns högskola. Research Reports 2023:1. ISBN 978-91-89504-47-9 (2023). https://sh.diva-portal.org/smash/record.jsf?pid=diva2:1768279

STA, Saving lives beyond 2020: the next steps, in *Recommendations of the Academic Expert Group for the 3rd Global Ministerial Conference on Road Safety* (Swedish Transport Administration: Borlänge, Sweden, 2020). https://trafikverket.diva-portal.org/smash/record.jsf?pid=diva2:1413366

STA, *Saving Lives Beyond 2025: Taking Further Steps: Recommendations of the Academic Expert Group for the 4th Global Ministerial Conference on Road Safety* (Swedish Transport Administration: Borlänge, Sweden, 2025). https://trafikverket.diva-portal.org/smash/record.jsf?pid=diva2:1936696

H. Wennberg, O. Dahlholm, Traffic safety implementation at local level in line with 2030 Agenda for Sustainable Development: synergies, challenges and enablers. Traffic Saf. Res. **5**, e000055 (2024). https://doi.org/10.55329/lgfl1703

H. Wennberg, P. Hyllenius Mattisson, Traffic safety footprint in sustainability practices and reporting—exploring the views of companies. Sustainability **16**(24), 10975 (2024). https://doi.org/10.3390/su162410975

G. Yannis, E. Michelaraki, Review of city-wide 30 km/h speed limit benefits in Europe. Sustainability **16**(11), 4382 (2024). https://doi.org/10.3390/su16114382

Chapter 10
Artificial Intelligence in the Skies: Reconciling European Regulation and Liability in Aviation

Federico Franchina

Abstract The chapter contends that Europe's AI Act effectively becomes aviation's new safety code by binding its "high-risk" obligations to EASA certification. It warns that safe adoption of flight-critical AI hinges on a three-way balance—tight ex ante regulation, fair liability rules, and protection of aviation's Just Culture—then sketches a two-track roadmap of rapid technical standards and slower liability reform to achieve it.

Keywords Aviation safety · EU AI act · Liability · Just culture · International cooperation · Quality management system · Human oversight

10.1 Aviation Safety and Evolution of Air Transport

The air transport liberalization and deregulation process initiated decades ago, immediately brought to light the "limits" of the physical and legal configuration of the European aviation industry.

The growth of civil aviation at the European level along with the massification of air transport that seems to have granted the passenger the legal status of a consumer (EU Regulation 261/2004), thereby transferring further obligations, mainly of a contractual nature, onto operators (companies, managers, service providers, etc.) is not always compatible with the risk and complexity factors that—in any case—continue to characterize air navigation.

Solutions such as automation and artificial intelligence (AI) appear to be necessitated by the constant growth of air transport, the challenges in attracting talented personnel, the limitations of airspace and airport capacity, as well as the evolving demands of passengers.

The integration of AI into safety-critical sectors represents one of the most profound technological and regulatory challenges of the twenty-first century.

F. Franchina (✉)
University of Messina, Messina, Italy
e-mail: franchinaf@unime.it

C. Bieder et al. (eds.), *Living with the New Safety Landscape II*,
SpringerBriefs in Safety Management, https://doi.org/10.1007/978-3-032-29343-5_10

Nowhere are the stakes higher than in aviation, an industry defined by its meticulous commitment to safety and reliability. Although its evolution and massification have occurred, it still represents a high-risk system.

As AI transitions from a theoretical concept to an operational reality in cockpits and air traffic control centers, the European Union has positioned itself at the vanguard of global governance. The EU's approach, culminating in the landmark Artificial Intelligence Act, is not merely a set of guidelines but a comprehensive legal architecture designed to manage risk, ensure ethical alignment, and, crucially, reshape the landscape of legal responsibility (European Union Aviation Safety Agency (EASA), 2023; EUROCONTROL, 2020).

10.2 The Genesis and Principles of the EU AI Act

The European Union's proactive approach to regulating AI aims to set a global standard for trustworthy AI, a phenomenon often termed the "Brussels effect." The legislative process for the EU AI Act commenced in April 2021 and concluded with its approval in April 2024, with the regulation set to take full effect 24 months after its publication.

The AI Act is based on a risk-based approach that categorizes AI systems into four tiers: unacceptable risk, high risk, limited risk, and minimal risk. Systems posing an "unacceptable risk," such as those using subliminal techniques to manipulate behavior or implementing social scoring, are prohibited outright. Systems classified as "high-risk," which are of paramount importance to aviation, are subjected to stringent legal obligations and a mandatory conformity assessment before they can be placed on the market.

For these high-risk systems, the Act mandates several foundational requirements grounded in the EU's "Ethical Guidelines on Trustworthy AI." These include principles of transparency and traceability, robust data governance to ensure high-quality and unbiased training data, and adherence to privacy laws such as the GDPR. The Act also imposes obligations for technical documentation, record-keeping, accuracy, robustness, and cybersecurity. A crucial pillar is the principle of human oversight, requiring that high-risk AI systems be designed for adequate supervision by natural persons.

10.3 The Human–AI Symbiosis: From "Human-Centered" Design to "Human Oversight"

AI fundamentally reconfigures the relationship between the human operator and the machine. The philosophy of human-centered design has long guided such complex industrial activities; however, the EU AI Act introduces a new, more rigid paradigm:

"human oversight." This shift from a design philosophy to a legally mandated function reframes the human not merely as a user, but as a legally designated supervisor, tasked with monitoring a system whose inner workings may be inherently opaque.

Human-centered AI is a design approach that prioritizes human needs and experiences, aiming to augment human capabilities with collaborative, intuitive, and explainable systems. In contrast, "human oversight," as mandated by Article 14 of the AI Act, is a specific and legally binding requirement. It obligates that high-risk AI systems are designed to allow humans to understand their capabilities, remain aware of automation bias, interpret outputs correctly, and ultimately decide to override or halt the system's operation.

This transition from the goal of a "human-centered approach" to the mandate of "human oversight" represents a fundamental paradigm shift, imposing a legal duty on a human to monitor and validate a potentially opaque system actively. This has significant consequences for liability, as a failure under a human oversight regime immediately raises questions about whether the human supervisor fulfilled their legal duty.

10.4 The Crucible of Liability: AI and the Restructuring of Legal Responsibility

Advanced AI challenges the foundations of legal liability, which for centuries have been built around human action, intent, and foreseeability (Risslandt, 1957).

The AI challenge has sparked a global debate over the most suitable liability regime, with the primary contenders being fault-based negligence and strict liability without fault.

- **Fault-Based Negligence:** This traditional approach requires a claimant to prove that the defendant breached a duty of care, causing the injury.
- **Strict Liability:** This regime holds a party liable for harm caused by their product regardless of fault. It offers clarity for victims and creates strong incentives for safety, but faces criticism for potentially having a "chilling effect" on innovation.

Traditional legal systems, both civil and criminal, are based on the assumption that a human agent makes a free and conscious choice, which produces an outcome for which they can be held accountable. This is the essence of culpability.

AI systems can "act" in ways not directly foreseen or intended by their creators or users. Their decision-making processes can be opaque (the "black box problem," referring to the inherent opacity of many advanced AI systems, whose decision-making pathways are not easily traceable or explainable), making it impossible or extremely difficult to trace a harmful outcome back to a specific human error.

To address these limitations, legal scholars have proposed a socio-technical perspective (Emanuilov, 2017; Diamantis, 2025). This approach reframes an AI accident not as a discrete technical fault, but as an outcome of the entire system in which the AI is embedded, including its design, the context of use, and surrounding organizational procedures. From this viewpoint, foreseeability is not about predicting a specific output but about foreseeing the *types* of harms that could arise from deploying a particular AI tool in a specific human environment, providing a more robust basis for assigning responsibility.

This also raises a crucial question: Can a system of responsibility based on human freedom and agency remain valid in the age of AI? The answer is that this system is being put to the test, to the point of creating a critical gap that makes it difficult, if not impossible, to attribute moral or legal responsibility to a specific human actor (designer, programmer, user) according to traditional rules of guilt and causation.

The main consequence is a potential denial of justice for victims, who may be unable to identify a responsible party and obtain compensation.

10.5 The Direct Integration of the AI Act into Aviation Safety Law

While the EU AI Act is a horizontal regulation with broad applicability, it is not merely a set of advisory principles for the aviation industry. Through specific and deliberate legislative mechanisms, the Act's most stringent requirements are directly and mandatorily integrated into the existing body of European aviation safety law as Article 108 of the AI Act mandates a direct modification to Article 17 of Regulation 2018/1139. This transforms the Act from a general framework into a binding, sector-specific regulatory instrument.

Aviation is explicitly identified as a high-risk domain under this framework. An AI system is considered high-risk if it is intended to be used as a safety component of a product or is itself a product covered by specific EU legislation listed in the annexes, which includes aviation. The rationale is clear: The failure or malfunction of such systems in the aviation environment can directly endanger the health, safety, and property of individuals.

Consequently, actors within the aviation supply chain are subject to a detailed set of obligations for any high-risk AI system they develop or use. The Act establishes a transparent chain of responsibility, imposing distinct duties on "providers" (the developers of the AI system) and "deployers" (users, such as Air Navigation Service Providers (ANSPs) or airlines).

Providers are responsible for the majority of the compliance burden. They must establish and maintain a robust risk management system throughout the AI's lifecycle. They are required to ensure the quality and governance of the data used for training and testing, create extensive technical documentation, and design the system for transparency, traceability, accuracy, robustness, and cybersecurity.

Deployers, in turn, have their own set of responsibilities. They must operate the system according to the provider's instructions for use. They must ensure that the individuals assigned to human oversight are competent, adequately trained, and have the necessary authority to perform their duties effectively. Deployers are also responsible for monitoring the system's operation for potential risks and reporting any serious incidents or malfunctions to the provider and relevant authorities.

At the heart of the obligations for high-risk systems is the concept of a Quality Management System (QMS), introduced by Article 17 of the Regulation.

This is not a simple checklist, but a comprehensive and well-documented governance framework that covers the entire lifecycle of the AI system.

The system must include written policies, procedures, and instructions covering regulatory compliance strategy, design, and development, testing and validation, risk management, incident reporting and communication.

10.6 Navigating the Trilemma: Regulation, Liability, and Culture

The core challenges identified throughout this work highlight the inherent tensions at the intersection of technology, law, and human factors.

- First, there is a fundamental conflict between the AI Act's regulatory demand for effective human oversight and the operational reality of opaque "black box" systems. The law mandates that humans must be able to monitor, interpret, and override AI meaningfully; yet, the very nature of advanced AI can make its decision-making processes inscrutable. This places the human overseer in a precarious position, legally accountable for a system they may not fully understand.
- Second, there is a collision course between the liability domain of assigning fault and Just Culture's foundational principle of limiting blame. The legal system's need to establish culpability to compensate victims risks undermining the psychological safety and trust required for open reporting, which underpins the learning process for systemic safety. Without careful safeguards, the pursuit of legal justice could inadvertently compromise the mechanisms of aviation safety.
- Third, there is the enduring tension between the drive for rapid innovation and the paramount need for safety and legal certainty. While regulatory sandboxes (e.g., structured and temporary legal derogations supervised by regulators to enable real-world experimentation with innovative business modes) are intended to spur development, ambiguities in the rules for real-world testing could stifle the final, critical stages of validation.

Navigating this trilemma is the central task for all stakeholders. It requires moving beyond siloed thinking and developing a holistic approach where technology design, regulatory policy, and organizational culture are co-developed in a mutually reinforcing manner (Center for Long-Term Cybersecurity (CLTC), 2020, Arbel 2024).

10.7 The Path Forward: Recommendations for a Coherent Ecosystem

To successfully chart a course for the safe and responsible integration of AI in aviation, a coordinated effort is required from all major actors within the ecosystem.

For Regulators (European Commission, EASA, National Authorities):

- Clarify Regulatory Ambiguities: It is essential to provide unambiguous guidance on the transition from "regulatory sandboxes" to "real-world" operational testing. This will give innovators the confidence to validate their systems in realistic environments without fear of premature regulatory penalties.
- Develop Coherent Standards: Technical standards for AI explainability and transparency must be co-developed in conjunction with legal standards for human oversight.

For the Aviation Industry (ANSPs, Airlines, Manufacturers, and Technology Providers):

- Invest in Human Factors: A massive investment in human factors research and training is non-negotiable. This must go beyond traditional skills and focus squarely on the new competencies required for Human–AI Teaming, including managing automation bias, interpreting probabilistic information, and developing mental models of AI behavior.
- Evolve Safety Management Systems (SMS): Existing SMS frameworks must be updated to explicitly account for AI as a core component of the socio-technical system. Risk assessments must analyze not just the AI's technical failure modes but also the potential failure modes of the human–AI interaction.
- Champion Data and Model Integrity: The industry should proactively lead the development of standards for high-quality, unbiased training data and for the robust validation and verification of AI models. This is not just a compliance exercise but a prerequisite for building trustworthy and reliable systems.
- Embrace Contractual Clarity: In the absence of complete legal certainty from the AILD, actors across the supply chain can mitigate risk by using clear contractual agreements that allocate responsibilities, warranties, and indemnities for AI performance and potential failures.

For Legal and Judicial Systems:

- Foster Judicial Expertise: The complexity of AI technology requires a judiciary that is equipped to handle these novel cases. Investment in judicial training and the use of expert witnesses will be crucial for ensuring that the new liability rules are applied consistently and correctly.

10.8 The Future of AI in Aviation: Beyond Current Frameworks

Comparing aviation governance with the challenges posed by AI yields nuanced yet valuable insights. The most important lesson from aviation precedent lies not in its specific rules, but in its governance process—a two-track approach that combines rapid technical standardization with slower, consensus-based evolution of accountability, all grounded in a proactive culture of safety management.

However, the limitations of this analogy are equally instructive. The nature of AI—general-purpose, software-based, rapidly evolving, and easily disseminable—makes a single global treaty (like that one signed in Chicago in 1944 that shaped the civil aviation) not only unlikely but perhaps undesirable. Today's geopolitical context is one of competition, not the post-war collaboration that made the 1944 Convention possible.

Therefore, the most promising path for AI governance appears to be a hybrid, polycentric model that:

1. Promotes international cooperation on key technical standards: Pursue a "fast track" to establish shared basic principles (e.g., on data quality, security, interoperability) to build a minimum level of global trust and confidence like it's happening at ICAO level with the adoption of standard and recommended practices.
2. Embraces adaptive and iterative liability regimes: Allows for a "slow track" where different jurisdictions can experiment with liability rules (such as the EU approach), learning and adapting over time.
3. Prioritizes procedural accountability: Supports the widespread adoption of ex ante governance mechanisms, such as the Quality Management System, as the most effective tool to fill the accountability gap and operationalize the principle of meaningful human control.

The current regulatory and legal frameworks are designed to manage the first wave of AI integration, where systems primarily function as decision-support tools or semi-autonomous agents under human supervision. However, the technological horizon extends far beyond this. The industry is already envisioning a future with single-pilot commercial operations, fully autonomous cargo flights, and AI-driven air traffic management where intelligent agents communicate and deconflict with one another seamlessly.

This future will present a new set of even more profound challenges. How will a fully autonomous aircraft be certified as safe? What legal personality, if any, will an autonomous agent possess? How will liability be allocated when an accident is caused by an interaction between two or more independent AI systems? And perhaps most importantly, how can public trust be built and maintained in an aviation ecosystem where human control becomes the exception rather than the rule?

Addressing these future questions will require another evolution in our legal, regulatory, and cultural thinking. The principles being established today—risk-based regulation, the primacy of human oversight, the need for transparency, and the balance between accountability and learning—will serve as the essential foundation for future development. The goal remains unchanged: to leverage the transformative power of artificial intelligence to enhance, rather than erode, the unparalleled safety, reliability, and public trust that define modern aviation.

Ultimately, the historical link between responsibility and human agency must be preserved. This will not happen by trying to force AI into old legal categories, but by building new regulatory frameworks that make humans responsible not so much for every single action of the machine, but for the design, implementation, and oversight of the systems they choose to create and bring into the world.

The journey is complex, but with a coherent and collaborative approach, it is possible to build a future where intelligent systems make the skies safer for all.

References

Agency Roadmaps, Guidance, and Reports

European Union Aviation Safety Agency (EASA), *EASA Artificial Intelligence Roadmap 2.0* (2023)
EUROCONTROL, *FLY AI Report: Demystifying and Accelerating AI in Aviation/ATM* (2020)

(Some) Academic and Research Papers

Y. Arbel, M. Tokson, A. Lin, Systemic regulation of artificial intelligence, 56 Ariz. St. L.J. (2024)
Center for Long-Term Cybersecurity (CLTC), *A Flight to Safety-Critical AI*. University of California, Berkeley (2020)
M.E. Diamantis, Reasonable AI: a negligence standard, 78 V and. L. Rev. 573 (2025)
I. Emanuilov, Autonomous systems in aviation: between product liability and innovation. Paper presented at SESAR innovation days 2017 (2017)
E.L. Risslandt, Artificial intelligence and law: stepping stones to a model of legal reasoning. Yale Law J. **99** (1957)

Chapter 11
AI and Offshore Safety Management

Ole Andreas Engen and Preben Hempel Lindøe

Abstract The chapter explores how artificial intelligence (AI) is reshaping governance, regulation, and professional practice in the Norwegian petroleum sector. It examines how AI affects political and regulatory cultures, reinforces existing power dynamics in safety management, and generates new forms of contestation. Set within a polycentric governance system that balances state authority, industry self-regulation, and international standards (Engen et al. 2023; Hansen and Antonsen 2024), the analysis shows how AI embeds within established regimes of control and accountability. This integration heightens tensions between transparency and secrecy and between self-regulation and prescriptive oversight. The chapter situates these dynamics within broader issues of algorithmic governance, highlighting frictions between AI-driven systems and a regulatory framework grounded in legal norms and professional discretion.

Keywords AI · Risk governance · Security · Power · Knowledge

11.1 Digitalization and AI

Artificial intelligence (AI) refers to computer systems capable of performing tasks that normally require human intelligence, such as learning from data, recognizing patterns, making decisions, and understanding language. As a disruptive technology, AI is transforming traditional industries by enabling new forms of automation, decision-making, and system optimization that were previously unachievable. However, AI does not operate in isolation. It is embedded within complex socio-technical systems that encompass not only hardware and software, but also people,

O. A. Engen (✉) · P. H. Lindøe
University of Stavanger, Stavanger, Norway
e-mail: ole.a.engen@uis.no

P. H. Lindøe
e-mail: preben.h.lindoe@uis.no

C. Bieder et al. (eds.), *Living with the New Safety Landscape II*,
SpringerBriefs in Safety Management, https://doi.org/10.1007/978-3-032-29343-5_11

organizations, and institutional frameworks. The intricate interactions among these components can generate new forms of vulnerability and failure.

The growing algorithmic complexity of AI systems, particularly in advanced domains such as machine learning, often produces outcomes that are unpredictable and sometimes inexplicable. A related challenge is the lack of transparency in AI systems, especially within deep learning models, whose internal operations and decision-making logic are difficult to interpret (Marr 2023). Together, these characteristics contribute to significant uncertainty surrounding the use of AI, creating situations where both outcomes and impacts are not only unknown but also continuously evolving.

Against this backdrop, the chapter addresses two central research questions:

- To what extent can a trust-based regime, built on mutual confidence, transparency, accountability, and collaboration, be maintained amid increasing digitalization, automation, and algorithmic governance? How does AI reshape knowledge production and interaction among actors within the offshore regime?
- How does the integration of AI into offshore risk management and regulatory practice redefine the relationship, and potential tensions, between "safety" and "security" perspectives?

The emergence of artificial intelligence (AI) in the offshore petroleum sector marks a profound shift in how risks are managed, how knowledge is produced, and how authority is distributed. In the Norwegian context—long governed by a trust-based, tripartite system that emphasizes collaboration between state, industry, and labor—AI introduces new dynamics that challenge established regulatory cultures and risk logics.

11.2 Application of AI in Norwegian Offshore Industry

AI is increasingly being applied in the Norwegian offshore industry to enhance safety, efficiency, and sustainability within energy production as oil and gas and offshore wind as well as in maritime operations (Lootz et al. 2025). Companies use AI-driven predictive maintenance systems on offshore platforms to identify equipment failures before they occur, reducing downtime and improving safety. Ringstad et al. (2025) illustrate how Equinor includes AI application in existing approaches to major accident safety barrier management. In offshore wind, companies are leveraging AI to monitor turbine performance and forecast energy production based on weather and ocean data. The maritime sector benefits from AI systems that support autonomous navigation, route optimization, and intelligent decision support for crew and operators. These applications are particularly valuable in the harsh and complex environments of the Norwegian continental shelf, where AI contributes to reducing operational risks, cutting emissions, and supporting the transition to more sustainable energy systems.

However, the rapid and transformative integration of AI also presents significant governance challenges, including issues related to transparency, accountability, safety validation, data ownership, and the changing roles of human operators. In the Norwegian context, such challenges must be addressed within the framework of the tripartite system, which brings together government authorities, industry actors, and labor organizations. This model, which has long underpinned offshore safety and regulatory development, will need to evolve to accommodate the complexities of AI, ensuring that innovation proceeds in a way that safeguards worker rights, maintains trust, and preserves the legitimacy of regulatory oversight in increasingly autonomous and data-driven environments.

11.3 The Norwegian Offshore Regime

Safety Management within the Norwegian Offshore Regime (NOR) was developed within the framework of a "Nordic Model," distinct from those elsewhere in Europe (Dølvik et al. 2015). Within this framework, a tripartite system based on mutual interests and emphasizing transparency, accountability, and collaboration among stakeholders was established (Engen et al. 2023). A key feature of the model is its in-house "Occupational Health and Safety Organization," which includes Working Environment Committees, independent Safety Deputies elected by the workforce, and occupational health and safety experts available for consultation internally or externally (Karlsen and Lindøe 2006).

Within five years, three major accidents became formative for the offshore safety regime. After a major fire on the Alpha platform (1975), the new Working Environment Act was extended to all permanent offshore installations. Two years later, the Bravo blowout caused a major oil spill, exposing environmental risks and prompting operators to develop internal safety systems. The Alexander Kielland disaster in 1980, which claimed 123 lives, became a key reference point leading to rules enforcing self-regulation. In these, the tripartite system of the Working Environment Act was adopted and implemented. However, implementation was not harmonious. Around the millennium, controversies about safety threatened the tripartite collaboration. Industry representatives claimed that Health, Safety, and Environment (HSE) conditions were good, whereas unions and the regulator argued that they had deteriorated, stating:

> "*... it seems that a culture has been established where breaches of regulations and procedures have been incorporated as normal practice and accepted.*"
>
> Former Director General of the Norwegian Petroleum Directorate, Gunnar Berge, in the annual report of the Norwegian Petroleum Directorate 2000.

During the first three decades, safety management in the Norwegian petroleum industry alternated between adversarial and cooperative modes of tripartite partnership (Rosness and Forseth 2014). Following political and regulatory intervention, a more cooperative climate emerged from the mid-2000s, with multiple arenas fostering learning (Lindøe 2024). The robustness of the Norwegian Offshore Regime lies not only in regulation but in the active engagement of all parties. Through dialogue arenas, the regime has enabled behavioral adaptation and informal interaction, strengthening its ability to include new actors and redefine roles in response to technological and socio-economic changes (Engen et al. 2023).

11.4 AI, Risk Governance, and the Offshore Regime: Transformations in Regulation, Power, and Knowledge

This section explores how artificial intelligence (AI) interacts with and transforms the institutional logics—both of safety and security—within the Norwegian Petroleum Regulatory Regime. It examines four interrelated themes: (1) the shifting regulatory and political cultures surrounding risk and security; (2) the redistribution of power and emergence of new interests across the safety-security nexus; (3) the resilience of trust-based governance amid increasing algorithmic opacity; and (4) the reconfiguration of knowledge production and actor relationships in hybrid human–AI systems. Together, these analyses illuminate not only the challenges AI introduces to existing regulatory and security arrangements, but also the possibilities for adaptation, coordination, and institutional renewal in response to technological change.

11.4.1 Political and Regulative Cultures: AI and the Changing Logics of Risk

AI's introduction into offshore operations is transforming political and regulatory cultures by disrupting the epistemological foundations of the Norwegian safety regime. Traditionally grounded in functional, risk-based regulation and mutual trust among stakeholders, the regime emphasized transparency, dialogue, and the use of expert human judgment. However, AI systems, particularly opaque and proprietary systems, complicate this landscape. These technologies challenge regulators' ability to trace decisions, evaluate safety justifications, and ensure accountable practices.

As detailed in Engen et al. (2026), the arrival of AI coincides with broader changes in risk governance following the implementation of the Norwegian Security Act.[1] The Act reframes the petroleum sector not just as a safety-critical domain but as a national

[1] The Norwegian Security Act (LOV-2018-06-01-24) entered into force in 2019, introducing a risk-based framework for protecting national security interests, including critical infrastructure and activities of vital importance to national sovereignty. The revised Norwegian Security Act (2019) gained renewed relevance after the war in Ukraine and the Nord Stream II explosions (2022), which underscored hybrid threats to offshore energy infrastructure and prompted closer integration of safety and security governance under Havtil.

security asset. This dual emphasis on safety and security produces new tensions. Safety logic is historically built on openness, dialogue, and learning—values that underpin trust-based regulatory models such as Norway's tripartite system in the petroleum sector. In contrast, security logic is premised on secrecy, centralization, and control, reflecting its roots in intelligence and national defense. As Jore et al. (2026) show, these two logics entail distinct ontologies and ethics: Safety assumes shared interests in preventing harm and therefore thrives on transparency, while security responds to intentional threats and often requires restricted information and hierarchical coordination.

Artificial intelligence (AI) technologies, especially when applied in surveillance, cybersecurity, or resilience planning, tend to align more naturally with the security logic. Their operation depends on data centralization, predictive control, and opacity (e.g., black-box algorithms and restricted data sharing), which may undermine the participatory and deliberative practices that safety cultures depend upon. The growing reliance on AI for risk detection and threat management thus risks reinforcing securitization trends that challenge long-standing norms of openness, trust, and collective learning central to safety governance. This shift introduces a new form of "algorithmic governance": one less reliant on visible compliance and more cantered on algorithmic oversight and pre-emptive risk detection, yet at the cost of transparency and participation.

11.4.2 Power Dynamics and Conflicting Interests

The introduction of AI also reveals and reshapes long-standing power dynamics in offshore safety governance as presented in the Norwegian tripartite system as presented above. Yet, AI threatens to displace this balance by creating new asymmetries of expertise and mechanism of control. Technological knowledge is becoming increasingly concentrated among system developers and vendors, who are often external actors with limited ties to the offshore safety tradition. This undermines the ability of regulators and employee representatives to assess, question, or intervene in AI-driven decisions.

Furthermore, AI reflects conflicting institutional logics. From a commercial perspective, AI is often promoted as a tool for efficiency and cost reduction. As one informant from the Norwegian Ocean Industry Authority (Havtil) noted, "The biggest risk with AI is not using it"—a statement that reveals the pressure to adopt new technologies for strategic gain, potentially at the expense of safety reflection and robust qualification. This "deployment imperative" can create organizational drift, where safety becomes a secondary concern. Meanwhile, regulators and union representatives may lack both the legal tools and technical insight to effectively counterbalance this trend.

Historically, Norwegian safety regulation has been shaped by a continuous balancing act between ambitions for risk reduction and the imperatives of commercial efficiency. These inherent tensions were traditionally managed through openness, dialogue, and negotiated compromise—anchored in institutional arenas such as the tripartite Safety Forum, where regulators, companies, and unions collectively defined what constituted an acceptable level of risk.[2] As Jore et al. (2026) demonstrate, however, the increasing securitization of industry practices in response to geopolitical instability and hybrid threats is transforming this balance.

Artificial intelligence (AI) and digital surveillance technologies intensify these dynamics by embedding security rationalities deep within everyday practices of risk governance. While such technologies promise predictive precision, operational efficiency, and enhanced control, they also centralize authority and obscure the basis for decision-making. The epistemic grounds on which risks are identified, assessed, and justified increasingly depend on algorithmic systems and classified information inaccessible to many of those affected.

Hence, at stake is not merely operational safety, but the epistemic authority to define what counts as legitimate knowledge and acceptable risk. When these judgements are delegated to data-driven or security-classified processes, the ethical foundations of safety governance—trust, transparency, and collective learning—are displaced by technocratic and security-oriented rationalities. This shift risks eroding the participatory ethos that has long underpinned the legitimacy and effectiveness of the Norwegian safety regime.

11.4.3 The Future of Trust-Based Governance

AI systems, particularly those that are opaque or proprietary, challenge the conditions under which trust can be sustained. Trust requires the ability to understand, challenge, and verify. When decision-making processes are embedded in non-transparent algorithms, these conditions are weakened. Moreover, the introduction of the Security Act and the growing focus on security-related risks have introduced institutional secrecy into the regulatory system. As Engen et al. (2026) shows, security risks require containment of information and controlled communication—a logic antithetical to the open sharing that has historically underpinned safety culture. The result is a hybrid governance regime in which trust must now be negotiated in the presence of opacity and control.

A representative of the Norwegian Ocean Industry Authority emphasized that sustaining trust in the era of digitalization and AI requires more than interpersonal confidence or institutional reputation. Trust, they suggested, must increasingly be

[2] The Safety Forum is the principal tripartite arena for dialogue between authorities, industry, and labour unions in the Norwegian petroleum sector. Established in 2001 and chaired by the Norwegian Ocean Industry Authority (Havtil), it serves as a collaborative platform for discussing safety, working environment, and regulatory issues.

supported by systems that make decisions transparent, accountable, and verifiable. This view reflects a broader recognition that regulatory trust can no longer rest solely on established relationships or assumptions of integrity but must be actively maintained through mechanisms that ensure traceability and explainability. A complete departure from trust-based regulation is neither inevitable nor desirable; rather, what is needed is a recalibrated model of trust that incorporates third-party auditing and grants stakeholders' meaningful access to AI logic and outcomes—conditions essential for maintaining legitimacy and shared ownership of risk governance.

Trust has historically been the cornerstone of Norway's offshore safety regime. It facilitates the free flow of information and knowledge necessary to achieve regulatory goals—such as safety reporting, risk analysis, and compliance oversight. Trust underpins the willingness of actors to share sensitive data and to engage in joint problem-solving, and it rests on the assumption of shared interests between government, industry, and labor. As Jore et al. (2026) emphasize, the tripartite system depends on this moral economy of trust to sustain legitimacy and ensure that rules and responsibilities can be negotiated rather than imposed.

However, trust is a fragile resource. It is continually exposed to structural pressures arising from dependency and power asymmetries between the state, corporations, and workers. These dynamics are especially evident in complex governance networks where political, legal, and administrative institutions interact with industry actors, research institutions, and trade unions. When the traditional safety logic based on transparency, participation, and shared learning meets a security logic rooted in secrecy and control, trust becomes both more valuable and more vulnerable.

11.4.4 AI, Knowledge Production, and Actor Interactions

AI also acts as a disruptive force in knowledge production. It generates new kinds of knowledge—statistical patterns, predictive maintenance indicators, and real-time anomaly detection—that were previously inaccessible. Yet, these outputs are often decontextualized, requiring careful interpretation by human actors. As noted in Ringstad et al. (2025), and Lootz et al. (2025), there is a risk of "de-skilling" among operators and HSE personnel who may come to over-rely on AI or lose confidence in their own judgment.

AI does not merely support human expertise; it competes with it. In doing so, it shifts the organizational hierarchy of knowledge, privileging data science over operational experience. This creates tension between traditional expert groups (engineers, operators, safety officers) and new actors (data analysts, system designers, AI vendors), especially when AI outputs conflict with established knowledge or fail to provide clear justifications.

To navigate this tension, it is crucial to embed AI into collaborative, interdisciplinary structures that ensure interpretability and human oversight. Knowledge must not only be produced, but also socially validated—especially in high-risk environments like offshore petroleum.

The increasing securitization of the petroleum sector, driven by geopolitical instability and digital transformation, is reshaping how information flows within this system. Security concerns often justify restrictions on openness, limiting what can be shared in collaborative arenas such as the Safety Forum. AI-based monitoring, predictive analytics, and classified cybersecurity protocols exacerbate these constraints. As Jore et al. (2026) show, workers and unions accept such measures to a degree, relying on trust in the authorities' intentions; yet this trust also acts as a buffer that conceals emerging inequalities in knowledge and influence.

Data sharing within and across sectors further illustrates these dilemmas. In safety management, effective coordination relies on reciprocal openness—actors must know enough about one another's systems to manage interdependencies. Within established value chains, such data sharing typically builds on cooperation and interdependence, where trust compensates for the absence of formal control. By contrast, in the new hybrid safety-security landscape, information is increasingly framed as a strategic asset to be controlled, not a collective good to be shared. Companies must now decide what to disclose and what to protect considering commercial, legal, and national security interests.

Historically developed tensions between safety ambitions and commercial efficiency are thus deepened by AI and the expanding security mandate. At stake is not only operational safety but also the epistemic authority to define what constitutes acceptable risk. As decision-making becomes more data-driven and security-classified, the ethical foundations of the safety regime—trust, transparency, and inclusivity—risk being displaced by technocratic and security-oriented rationalities. Safeguarding trust in this new environment requires governance arrangements that balance the legitimate need for confidentiality with continued opportunities for joint learning, participation, and accountability.

11.5 Final Remarks

Digitalization and artificial intelligence (AI) are reshaping the foundations of risk management in high-tech and high-risk industries such as the petroleum sector. Experiences from the Norwegian Ocean Industry Directorate (Havtil) illustrate how AI challenges established regulatory regimes and control systems—technically, organizationally, and normatively. AI systems are often less reliable than traditional technologies, highly complex, and opaque. This introduces new forms of risk, particularly predictive and systemic risks, making it difficult for both regulators and industry to anticipate and manage adverse events.

A key concern is that the safety perspective may weaken as risk assessments become dominated by data security, privacy, economic, and commercial considerations. As digital infrastructures expand and AI-driven systems become integral to monitoring, maintenance, and decision support, information itself turns into a strategic resource. The boundaries between safety, security, and competitiveness blur. Decisions about disclosure are increasingly guided not by collective safety needs but by ownership, confidentiality, and liability. This alters the epistemic conditions under which risks are identified, shared, and acted upon.

When AI systems and cybersecurity architectures classify information as sensitive or proprietary, data vital for learning and prevention may become inaccessible to regulators, unions, and workers. The transparency and deliberation that traditionally sustain safety culture risk are being replaced by selective disclosure and institutionalized secrecy. Over time, trust may shift from being grounded in participatory practice to becoming an expectation of passive compliance, as prescriptive rules leave less room for discretion and must be strictly followed. Safety is thus reframed not as a shared ethical commitment to prevent harm, but as a subset of organizational risk management shaped by technocratic and commercial rationalities.

In sum, experiences from the petroleum sector show that AI unfolds in a domain where small failures can have major consequences, and where the interaction between humans, systems, and organizations is critical. AI raises not only technical and legal issues, but also fundamental challenges related to governance, accountability, and trust in modern risk societies. Its introduction represents more than a technological shift; it signifies a transformation in governance, culture, and power. AI reshapes how risks are perceived, how decisions are made, and who holds authority. While the Norwegian system's strengths—trust, dialogue, and adaptability—remain vital, they are under pressure from opaque systems, security imperatives, and shifting institutional priorities.

Maintaining a resilient and democratic form of risk governance in this context requires active efforts to realign regulatory tools, distribute epistemic authority, and preserve trust-based collaboration even in an era of algorithmic reasoning and heightened security concerns.

Ethical Approval This study is based on data collected through interviews with representatives of the Norwegian Ocean Industry Authority (Havtil) in 2025 as part of the research project INTERSECT—New Geopolitics and the Interaction between Safety and Security in Petroleum Risk Governance. The data collection was approved by SIKT—the Norwegian Agency for Shared Services in Education and Research (reference no. 904592), which operates within the framework of the Norwegian Act relating to Ethics and Integrity in Research and the system of the National Committees for Research Ethics. All research activities were conducted in accordance with established Norwegian ethical guidelines, including requirements for informed consent, confidentiality, and secure data handling, and all interview data have been fully anonymized. Norwegian ethical regulations are aligned with the principles of the Declaration of Helsinki, ensuring voluntary participation and the protection of participants' rights, integrity, and privacy.

References

J.E. Dølvik, T. Fløtten, J.M. Hippe, B. Jordfald, *The Nordic model towards 2030: A new chapter?* Fafo Institute for Labour and Social Research (2015)

O.A. Engen, P.H. Lindøe, G.S. Braut, Coping with different system logics of standardization in regulatory regimes: Norwegian offshore experience. Saf. Sci. **161**, 106079 (2023). https://doi.org/10.1016/j.ssci.2023.106079

O.A. Engen, S.H. Jore, S.T. Hansen, Polycentric risk governance across safety and security risk logics (2026)

S.T. Hansen, S. Antonsen, Taking connectedness seriously: a research agenda for holistic safety and security risk governance. Saf. Sci. **173**, 106436 (2024). https://doi.org/10.1016/j.ssci.2024.106436

S.H. Jore, O.A. Engen, S.T. Hansen, Ø. Midttun, Ethical dilemmas when safety meets security, in *Ethics in the Doing of Technoscience. Navigating Professional Frameworks and Organizational Muteness*. ed. by J. Hayes, S. Maslen (Palgrave Macmillan, 2026). https://doi.org/10.1007/978-3-032-20002-0_9

J.E. Karlsen, P.H. Lindøe, The Nordic Osh Model at a Turning Point? Policy Pract. Health Saf. **4**(1), 17–30 (2006)

P.H. Lindøe, The Tripartite System: a key in polycentric risk governance: lessons from Norwegian Offshore Industry, in *The Regulator-Regulatee Relationship in High-Hazard Industry Sectors. New Actors and New Viewpoints in a Conservative Lanscape*. (Springer, 2024), pp.45–54

E. Lootz, L.I. Vestly Bergh, B. Heide, Artificial intelligence and uncertainty in a high-hazard industry: the role of risk management and regulation, in *Proceedings of the 35th European Safety and Reliability and the 33rd Society for Risk Analysis Europe Conference*. ed. by E.B. Abrahamsen, T. Aven, F. Bouder, R. Flage, M. Ylönen (Research Publishing, 2025), pp.1420–1427. https://doi.org/10.3850/978-981-94-3281-3_ESREL-SRA-E2025-P2957-cd

B. Marr, *Artificial Intelligence in Practice: How 50 Successful Companies Used AI and Machine Learning to Solve Problems* (Wiley, 2023)

A.J. Ringstad, A. Fernandes, J.T. Ludvigsen, Artificial intelligence and major accident risk: towards a framework for systemic safety risk assessment. In E. B. Abrahamsen, T. Aven, F. Bouder, R. Flage, M. Ylönen (Eds.), *Proceedings of the 35th European Safety and Reliability & the 33rd Society for Risk Analysis Europe Conference* (Research Publishing, 2025), pp. 1412–1419. https://doi.org/10.3850/978-981-94-3281-3_ESREL-SRA-E2025-P0734-cd

R. Rosness, U. Forseth, Learning and adapting in a high-risk industry: the Norwegian petroleum sector. Saf. Sci. **68**, 11–23 (2014). https://doi.org/10.1016/j.ssci.2014.03.003

Part III
Conclusion

Chapter 12
Regimes of Fractured Objectivity: On the Limits of Technological Risk Governance

John Downer

Abstract This chapter argues that modern risk governance rests on a useful fiction: that technological safety can be rendered as objective fact. Within a "regime of functional objectivity," institutions translate unruly hazards into standardized metrics that appear auditable, comparable, and decisive. This fiction works tolerably well because the uncertainties of most technologies can be bounded enough for regulation and coordination. The chapter contends, however, that this regime has an epistemic boundary, past which the assumptions that support objectivity claims can no longer be safely ignored. Beyond this threshold lies a "regime of fractured objectivity," where extreme stakes and sparse empirical feedback demand implausible certainty, small modeling choices have outsized effects on outcomes, and routine heuristics of expertise lose traction.

Keywords Technological risk · Risk governance · Civic epistemology · Objectivity · Nuclear safety · Epistemic uncertainty · Science and Technology Studies (STS)

Van·tas·ner Dan·ger Me·rid·i·an

/vanˈtasnərˈ dānjər məˈridēən/

noun

1. The point or line after which danger to your mission increases exponentially. Often used to demarcate conditions of encroaching and grave danger.[1]

---

[1] Conrad et al. (2018).

J. Downer (✉)
School of Sociology, Politics and International Studies (SPAIS), University of Bristol, Bristol, UK
e-mail: john.downer@bristol.ac.uk

C. Bieder et al. (eds.), *Living with the New Safety Landscape II*,
SpringerBriefs in Safety Management, https://doi.org/10.1007/978-3-032-29343-5_12

12.1 Introduction

The laws by which we harness physics often hold only inside specific envelopes. Cross a threshold and familiar rules stop giving the right answers; one way of thinking gives way to another.

Take aerodynamics. At low speeds, eighteenth-century theoretical tools—Bernoulli, Euler—accurately predict how air flows over a body. Push that body toward the speed of sound, however, and those tools start to slip. Around Mach 1, pockets of the airflow go supersonic while others lag. This marks a boundary into what is known as the "transonic regime," marked by shocks, sharp jumps in pressure and temperature, a spike in drag, and new stability behaviors. To operate inside this new envelope engineers must drop their old models and switch to new tools: Navier–Stokes equations, Rankine–Hugoniot shock jumps, and more.

Beyond this boundary lie others. If the engineers could propel that body (dramatically) faster still, it would eventually hit a different frontier, this time in relation to space–time itself. At about a tenth the speed of light, Newton's laws of motion would begin to falter: time would slow measurably, length would noticeably contract, and momentum would swell in ways they never imagined. Beyond this boundary, the "Newtonian regime" would end and engineers would need to invoke special relativity in their calculations. If our improbably resourceful engineers could (again, dramatically) shrink the body instead,[2] they would meet a different boundary: one where classical physics gave way to the quantum regime. At this point, things would get a little peculiar. Precise trajectories would stop making sense, and they would have to start working with probabilities and quantized states.

There are many such boundaries in physics and engineering. Some are relatively crisp and well-defined. The Mach threshold, for instance, marks a mostly constant and unmistakable transition into the transonic regime. More often, however, these boundaries are fuzzy because the regimes they separate are conceptual conveniences rather than physical discontinuities. Relativity and quantum mechanics always apply, but, under most conditions, it is more practical to act as if they do not until the limits of that pretense start to become a practical issue.

This chapter proposes an analogous regime boundary in the study and practice of technological safety. It argues that, like the limits of Newtonian or quantum regimes, our systems of safety reasoning also operate within envelopes that eventually fail. An indistinct yet critical threshold exists where familiar heuristics and analytical tools begin to lose purchase and new forms of understanding become necessary. Beyond this point established frameworks for conceptualizing technological safety begin to falter because the assumptions about authority, confidence, and institutional capacity on which they rest no longer hold. The chapter explains the logic of this regime boundary and explores the implications of operating beyond it.

The argument unfolds in four sections. Section 12.2 maps the civic epistemology that has long shaped how states know and govern technological risk. Section 12.3

[2] Technically, to the point where the de Broglie wavelengths of its particles rival the dimensions of the space confining them. Whatever that means.

shows how this civic epistemology rests on a useful fiction: that safety can be treated as an objective property of artifacts. And it argues that this fiction is useful because of certain conditions that constitute what I call "the regime of functional objectivity." Section 12.4 examines how those conditions can break down, undermining the usefulness of objectivity claims and the civic epistemology they sustain. It outlines an epistemic boundary I refer to as "Vantasner," beyond which the heuristics of expertise lose traction, the machinery of risk governance starts to fail, and claims of safety outrun what can meaningfully be known. Finally, Sect. 12.5 considers what governance might look like beyond that boundary, in what I call "the regime of fractured objectivity," where assurance gives way to uncertainty, and safety work becomes governance at, and sometimes beyond, the limits of knowability.

12.2 Civic Epistemologies of Technological Risk

Over the past century, states have assembled an elaborate and broadly coherent repertoire for managing technological risks: analytical tools, institutions, and procedural norms through which hazards are made "knowable" and therefore governable. Specialized agencies and regulatory bodies translate complex technical uncertainties into administrable objects via metrics, models, and categories that render uncertain futures legible to bureaucratic reason. In doing so, they convert indeterminate hazards into standardized quantities that can be compared, ranked, and managed within familiar decision frameworks.

Over time, this repertoire has become deeply institutionalized, providing states with a durable architecture of technological risk governance: an assemblage of expertise, authority, and practices through which the hazards of innovation are managed. Social scientists from a range of disciplines have spent decades exploring the many braided strands of this assemblage, unpacking the psychological, organizational, social, and normative commitments on which our technological risk governance rests (e.g., Silbey 2009).

These structures by which states govern technological risk, together with the scholarship that examines them, both reflect and co-produce what Jasanoff (2005) calls our civic epistemology of technological risk. This can be thought of as the patterned habits of reasoning and trust through which societies decide what counts as expert knowledge and how it should inform governance: an interlocking set of assumptions about how risk is known, managed, and justified. Sustained by cultural and institutional routines, this order shapes judgments of responsibility, credibility, and accountability in the management of technology, while delineating the boundaries of expert authority and determining how that expertise feeds into policy and practice.

A useful way to trace the contours of this civic epistemology of technological risk is to contrast it with that of economic or financial risk. Risks in both domains are governed as matters of public concern, shaped by expert input, institutional oversight, and political decision. Yet each follows a distinct pattern of public reasoning: mobilizing different tools and heuristics, empowering different organizations, and legitimating different kinds of claims. The standing of experts varies accordingly, as do the expectations placed on them, the measures of success, the burdens of proof, and the modes of accountability.

Economic risk is treated as inherently uncertain, and expertise pertaining to it as indispensable but invariably provisional. On this topic, expert pluralism is expected: rival schools with rival forecasts, and no single voice presumed final. Policymakers in this context often make choices about which experts to heed and how to interpret their advice. Non-expert perspectives are built into decision-making processes, and open contestation over data and interpretation is routine. And when losses occur, inquiries may follow but economists themselves are rarely held to account.[3]

Technological risk, by contrast, is treated as more definitive and controllable and expertise pertaining to it as more determinative. Consensus is expected in this context, with policymakers typically deferring to designated technical authorities, who are expected to fix probabilities and thresholds with standardized tests and models. Non-expert input is channeled through formal consultation and compliance processes, and disputes are referred back to procedure or expert committees. And when losses occur, inquiries tend to focus on failures of testing, certification, or design, with experts (or expert organizations) often facing professional sanction or legal liability.[4]

This dichotomy is an ideal–typical simplification, to be sure, but it is a useful one. Different societies frame these risks in distinct ways, and closer examination would reveal fine-grained choices shaped by cultural assumptions and institutional habits (indeed, this is Jasanoff's point in describing civic epistemologies). But taken in the round, the caricature captures stable and recognizable patterns in how modern states approach different types of risk. Finance ministers may push through a budget in defiance of gloomy forecasts from expert bodies, but transport ministers cannot launch a high-speed train until regulators sign off on every brake, signal, and software system.

This difference persists because we operate in an epistemological environment where it is, at least, functional, even if arguably not always ideal, to treat technological risk claims as if they are objective facts; in what I will call a *regime of functional objectivity*.[5]

[3] The 2022 UK "gilt crisis" illustrates the pattern. A mini-budget without an OBR forecast sparked deficit and inflation worries; economists disagreed on the scale and timing of the risk; gilt yields jumped, LDI funds came under stress, the Bank of England launched a temporary gilt-buying operation, and the government-reversed course. Political heads rolled; those of economists did not (Pinter 2023).

[4] The 2017 Grenfell Tower fire is illustrative here. Dangerously flammable cladding materials were certified as compliant under prevailing testing and approval regimes, contributing to a fire that killed seventy-two people. The subsequent public inquiry exposed deep flaws in testing standards and regulatory oversight, and accountability ultimately settled on technical authorities (Hodkinson 2018).

[5] I use "regime" here in the sense of a stable configuration of practices and assumptions.

12.3 Regimes of Functional Objectivity

Underpinning our civic epistemology of technological risk is a realist presumption: that "risk" is a knowable, quantifiable property of technological systems, its uncertainties disciplined through tests, models, and experiments designed to yield reproducible facts (Daston and Galison 2007; Porter 1995).

Within this order, legitimacy arises within what Porter (1995) calls the "moral economy of quantification," in which credibility is secured through the visible performance of methodological rigor. In this context, the ideal of technoscientific objectivity functions both as a means of coordination and as a claim to legitimacy: grounding governance in the capacity to translate ambiguous hazards into standardized and auditable referents that can circulate across institutions and sustain shared judgments of what counts as "safe enough."

In this way, objectivity underwrites a broader political order in which the technical validation of safety substitutes for its social authorization (Jasanoff 2005). Regulators, courts, insurers, and firms govern through ostensibly neutral procedures, using metrics and documentation as tokens of accountability. The authority of expertise thus rests not simply on technical proficiency, but on the institutionalized belief that safety, once properly measured, can be rendered "knowable." Absent this conviction, the machinery of technological risk governance would collapse under its own ambivalence (Beck 1992).

I call this claim to technoscientific objectivity an *ideal* because philosophers of science and STS scholars have long known it to be, on some level, a fiction. A core tenet of those disciplines is that no scientific or engineering claim is ever perfectly objective in a strict epistemological sense because such knowledge claims all rely on subjective judgments: about the representativeness of tests and experiments, for instance, or about the skill of the testers and experimenters (Latour 1987; Kuhn 1996 [1962]). This means that any statement about a technology's properties, especially its risks, always stands on a hidden scaffolding of assumptions.[6] Safety assessments translate unruly realities into stylized abstractions, and the safety ratings, certifications, and benchmark scores they produce are only as good as the assumptions implicit in those translations. This is to say that the "facts" of technological risk are "made" as much as they are "discovered," and that every safety assurance retains a residue of subjectivity and uncertainty, however well concealed (Wynne 1988; Downer 2024).[7]

---

[6] As part of certifying the safety of jetliners, for instance, regulators fire chickens at engines to simulate bird strikes. The procedure has an air of precision, but its results hinge on quietly contested choices about which species, what speed, how many, and under what conditions (Downer 2007).

[7] There is an interesting question about the extent to which engineers themselves subscribe to ideals of objectivity and certainty. Those closest to the practical work of designing and testing complex systems are often acutely aware of their uncertainties (MacKenzie 1998). Yet the discipline also carries a long tradition of idealized deductive reasoning, and with it, arguably, an overconfidence in calculation, especially where results are not tested against experience (Petroski 1992; Layton 1976).

Scholars have long debated the practical merits of sustaining this fiction. Some argue that performances of objectivity have long underwritten the public authority of science (Shapin 1994; Porter 1995), and contend that managed displays of certainty serve an important social function, especially amid rising vaccine skepticism and related enthusiasms (Oreskes 2019). Others argue that concealing expert uncertainties from non-technical audiences is inherently problematic for our risk governance. Jasanoff (2003), for example, calls the institutional processes by which risk is made to look objectively quantifiable "technologies of hubris." She argues that by allowing experts and institutions to claim an unrealistic degree of control, they blind us to uncertainty and ambiguity, suppress political debate, and ill-equip us for challenges that lie beyond the categories they themselves impose. She calls instead for "technologies of humility": institutional practices that confront, rather than conceal, the limits of prediction and control and make technical uncertainty a site of reflection and inclusion.

But while we can question whether it is optimal and/or democratic to mask uncertainties, it clearly allows for at least a minimally functional relationship to technological risk. A realist presumption of objectivity and the certainty it seems to promise may both be illusory, but they serve as useful heuristics for most technology governance. This is largely because, as Constant (1999) observes, engineering is less about ontological truth than about whether systems behave as expected under foreseeable conditions. Policymakers rarely need perfect knowledge of the systems they govern. They need approximations to hold, assumptions to stay within safe bounds, and margins of error to remain acceptable for the task at hand. And experts can meet this threshold in most contexts.

This is all to say that we think about technological risk in the manner that we do, and organize it the way we do, because on some level it *works*. The epistemological uncertainties of "knowing" most systems are manageable enough to design, operate, and regulate them with justified confidence that their risks are under control. The fact that we treat technological risk differently than economic risk reflects meaningfully different affordances and constraints. John Kenneth Galbraith's famous (and possibly apocryphal) quip about economic forecasting existing chiefly to make astrology look respectable finds no parallel in technology assessment. When engineers declare a bridge too weak to stand or a rocket too underpowered to fly, policymakers ignore them at their peril.

Most of the social scientific scholarship around technological safety benefits from the same conceit. Treating its technical dimensions as stable, knowable, and largely unproblematic licenses scholars to focus on its social processes without opening esoteric questions about the systems themselves. In doing so, it performs what Gieryn (1983) calls "boundary-work": allowing technical experts and social scientists to demarcate their areas of competence, and thereby coexist without encroaching on each other's professional jurisdictions.[8]

---

[8] This arrangement sits comfortably with funding regimes and research cultures that valorize interdisciplinarity but don't question disciplinary authority.

Treating the technical as objectively knowable also underwrites a broader deference to technological ambition. Few safety scholars maintain that perfect safety is achievable, but most proceed from the premise that "safe enough" can be secured through improved design, foresight, or organizational learning.[9] So it is that even accounts that accentuate the limits of formal assurance, such as Hollnagel's (2014) Safety-II, usually accept that technological risks are fundamentally manageable, and thereby preserve the field's accommodation with engineering expertise.

For all its utility to practitioners, decision-makers, and scholars, however, the regime of functional objectivity has limits. Push the safety demands on technological systems far enough and a boundary appears where differences of *degree* shade into differences of *kind*. Beyond this point, the pragmatic pretense of objectivity ceases to be useful and begins to mislead. The heuristics that ordinarily sustain expertise and authority become dysfunctional, and the conventional practices of risk governance start to falter.

This is a consequential boundary, deserving of an authoritative label. We should probably call it the "Epistemic Limit" or the "Ontological Uncertainty Threshold," or something. For the purposes of this chapter, however, I will refer to it as *Vantasner*.[10]

12.4 Vantasner

Let us say that Vantasner marks the point beyond which the inherent epistemological uncertainties of all technical knowledge can no longer safely be ignored in the practice of managing technological risk. A threshold where concerns about the formal limits of proof, which would usually be of interest only to academic observers, start to undermine the capability of technical experts and the credibility of their assertions.

Risk management practices often extend beyond this boundary due to characteristics of the technologies they address. One such characteristic is the raw complexity and novelty of the system in question. Some technologies are just more difficult to assess than others. They may rely on untested materials, operate in domains where the science is unsettled, or lack the real-world track record needed for calibration. In such cases, even modest safety claims drift quickly into speculation. Artificial intelligence is a useful illustration. A recent and rapidly evolving innovation with diffuse and potentially radical affordances, it has no meaningful service record, no comparable reference class, no clear explanation for why it produces particular outputs, and no definitive list of the hazards it may entail. Under these conditions, the uncertainties of its risk assessments multiply rapidly. And the (increasingly fashionable)

[9] Perrow (1999) is a notable exception here.

[10] The term is borrowed from *Patriot*, an underappreciated television dramedy, ostensibly about intelligence agents wrestling with Iran, but really more concerned with the pathologies of masculine stoicism. The show defines the term on-screen, in the manner reproduced at the top of this chapter: as a risk tipping point, beyond which hazards begin to become unmanageable. It seemed apposite.

claim that AI risks can be managed with sufficiently rigorous regulation is unlikely to be realistic (Blumenthal and Hawley 2023; Narayanan and Kapoor 2024).

Novel technologies tend to become better understood over time, of course, but a more stable and revealing dimension of Vantasner appears when the potential hazards of a technology become very extreme. This is because extreme hazards demand extreme certainty. If the consequence of a particular system failing might be the end of human civilization, we have to be very confident that it won't fail.[11] But when safety claims are stretched to the limits implied by one-in-a-billion errors, and mean-times-to-failure measured in millions of years, the ordinary background assumptions sustaining them can no longer be treated as negligible. In this context, the small but inevitable uncertainties of normal technological practice take on growing significance, exposing the limits of what can meaningfully be measured or verified.

A useful way to see the shape of this very generalizable problem is in relation to forensic DNA testing. In court, such tests are often said to be 99.9999999% reliable: an error rate of one in a billion. This is the statistical chance of two strangers sharing a genetic profile. But at that scale, the institutional assumptions that make the test possible—uncontaminated samples, calibrated machines, intact chains of custody—can no longer be taken for granted because the likelihood of contamination, equipment malfunction, or procedural error dwarfs the statistical probability of two strangers sharing a profile. Run a billion cases and you would see contamination, malfunction, and misconduct with regularity. So it is that a positive DNA test cannot come close to proving guilt to a one-in-a-billion degree of certainty (Lynch and Cole 2008).[12]

Claims about reactor safety offer a more consequential illustration. Experts routinely "prove," at enormous effort and expense, that the probability of a new reactor catastrophically melting down is no more than once in every million years of operation.[13] But reactors are sprawling socio-technical systems: they run at physical extremes, depend on unconventional materials, and are expected to function for decades. To sustain such remarkable failure claims, therefore, risk assessments must presume an extraordinary degree of confidence in an extraordinary number of

[11] Technically, failure severity is only one driver of extreme reliability demands; scale also matters. This is simply to say that a hazard tolerable for a single system becomes unacceptable when multiplied across many instances, since each must be safer to keep the aggregate risk in check. This is why jetliners and reactors are held to similarly unforgiving reliability standards, despite their very different hazards: we build far more jetliners than we do reactors (Downer 2017, 2024).

[12] This logical error, wherein a random-match probability is conflated with the posterior probability of innocence, is sometimes called "the prosecutor's fallacy" (Thompson and Schumann 1987).

[13] In a declaration to a UK regulator, for instance, Areva, a French reactor manufacturer, asserted that the likelihood of a meltdown in its "EPR" reactor was of the order of one incident every 1.6 million years of operation (in Ramana 2011).

assumptions. Experts mitigate this problem with system-level measures like redundancy, but these measures are themselves laden with assumptions.[14] At some point, the math gives way, and what remains is a performance of certainty, not its proof.

Reactor safety assurances are premised on such unrealistic failure calculations because catastrophic meltdowns are intolerable to a degree that few really appreciate. Their death tolls are battlegrounds of wildly varying numbers,[15] but the picture remains clear if we focus solely on their (less contested) monetary costs. A 2016 study of Chernobyl estimated its 30-year cost to be roughly $700 billion (Samet and Seo 2016). Studies of Fukushima have put its cost at over $500 billion (Behling et al. 2019). And the latter accident was almost far worse. Numerous authoritative studies of Fukushima, including the Japanese government's own official report, have concluded that the country came startlingly close to having to evacuate (long-term) its capital city (NAIIC 2012; Lochbaum et al. 2014). In 2011, the population of Tokyo and its surrounding prefectures was around 36 million people; the city was, and remains, a key pillar of the global economy. No state would pursue an energy infrastructure, or any technology, that it believed could realistically incur such costs with non-negligible probability. Reactors are only able to exist, politically and bureaucratically, therefore, because experts "prove" that the chance of them failing catastrophically is too low to merit it even being discussed (akin to other far-fetched hypotheticals like giant meteorite impacts or extra-terrestrial invasions) (Rip 1986: 7–9; Wellock 2017).[16] Hence, the need for reliability claims that are impossible to reconcile with the necessarily imperfect practice of actually assessing reliability (Downer and Ramana 2021).[17]

Beyond the unrealistic certainty they demand, systems with catastrophic failure conditions further complicate safety assessments by denying experts the insights they would usually glean from experience. In most circumstances, experts can refine their designs and assessments by drawing on actuarial evidence of how technologies actually perform in the world. (Medical regulators, for instance, recalibrate reliability claims using post-market performance data.) Observers, meanwhile, can use that experience to calibrate the authority they afford to the experts and their claims. When failures are intolerable, however, that feedback loop breaks down, as there

---

[14] The Fukushima disaster illustrated this. Erroneous assumptions about tsunamis led planners to place all the plant's redundant emergency generators below ground, where they flooded simultaneously (Lochbaum et al. 2014).

[15] The International Atomic Energy Agency (IAEA 2005) puts Chernobyl's death toll at "no more than 4,000 deaths," while a comprehensive study by Soviet scientists concludes it is over a million (Yablokov 2009). Between these extremes lies a wide spectrum of other seemingly authoritative figures.

[16] The fact that reactor catastrophes demonstrably, empirically, occur far more often than giant meteorite strikes or extra-terrestrial invasions, and that this doesn't seem to undermine the institutional credibility of reactor safety calculations, speaks to the authority of our civic epistemology of technological risk (Downer 2014).

[17] It is worth noting that reliability claims are essentially 'confidence' claims (about the likelihood of failure). This is apparent if we consider confidence bounds on reliability claims. It makes little sense, for instance, to say we are 99% confident a system will not fail over a given period, but only 25% confident that that number is correct.

can be no meaningful record from which to learn (and the cost of creating one is too high). This means that experts must rely much more heavily on untested assumptions. It also means that publics cannot calibrate their trust against lived experience. The result is a double bind: claims reach beyond what methods can justify even as the overreach grows harder to detect.

Another way that catastrophic failure conditions challenge safety claims is by pulling in contingencies that would normally be dismissed as irrelevant. We saw above that all risk analyses rest on assumptions that become problematic as their claims grow more extreme. But such analyses also *ignore* a great deal; not just because they are unlikely, but because they are judged to be outside the purview of technological risk management. When regulators certify the safety of a new hatchback, for example, they do not ask whether the car might be swallowed by an earthquake, consumed in a wildfire, swept away by a tsunami, struck by a missile, or undone by any number of other misfortunes. Such eventualities fall outside the scope of automotive risk assessment. None is inherently implausible, in the sense that each has surely claimed a hatchback at some time or another, but they are irrelevant to the policy judgments that automotive safety assessments are intended to inform. If someone is firing missiles at your hatchbacks, you have larger problems than vehicle regulation.

As the potential hazards of a technology grow, however, managing its risks means accounting for a much broader range of possibilities. If an earthquake at your reactor site could cost you your capital city, it cannot reasonably be ignored in the assessments that inform your policymaking. This is why reactor risk assessments explicitly model earthquakes, wildfires, tsunamis, and other "Acts of God" that technology risk assessments usually ignore.[18] But no assessment can anticipate every contingency. In the late 1970s, for instance, when the Zaporizhzhia plant was approved for construction in then-Soviet Ukraine, no one considered that it might one day sit on the frontline of a war with Russia. That possibility fell outside its design basis, but it should not have. If reactor risk assessments are to exclude catastrophic scenarios from policy discussion—as they must for the technology to be viable—then we must understand them as offering accurate probabilities of catastrophe from *any* cause.

Much more could be said about the conditions that push technologies beyond the limits of conventional risk analysis, but the central argument should now be clear. Certain properties of some technological systems—notably, though not exclusively, their catastrophic potential—can overstretch the credibility of the processes by which their safety can be known. At this boundary, the premises of conventional risk governance begin to erode. The work becomes a more explicitly speculative exercise, demanding that analysts confront a vast horizon of possibilities, imagining threats that may emerge decades ahead, many lying far beyond the reach of conventional technical expertise. It also requires they assign implausibly precise probabilities to those threats, projecting a degree of confidence that would be misleading even in mundane circumstances. And it compels them to do all this without the benefit of real-world experience to test or correct their claims.

[18] Albeit not always successfully, as Fukushima testifies.

Beyond this point begins what I will call *the regime of fractured objectivity*. Here, the kinds of ambiguities that interest scholars of socio-technical knowledge start to have real, tangible policy implications, and cannot be ignored. The fiction that technological risks can be accurately measured ceases to be useful, and the civic epistemology built on that fiction begins to unravel. Technical assessments lose their authority and no longer provide a stable basis for coordination, accountability, or control. As their credibility erodes, so too do the rituals it supported: verification, certification, compliance. Institutionalized deference to technical assertions stops being pragmatic, and the conviction that safety is always attainable starts to look dangerous rather than aspirational.

Reckoning with technologies in this regime means reshaping our civic epistemology of technological risk to grasp the nettle of technical uncertainty.

12.5 Regimes of Fractured Objectivity

For scholars, publics, and policymakers alike, governing technological risk in the regime of fractured objectivity poses complex questions with no perfect answers. It means rethinking our civic epistemology of technological risk in ways that accommodate fundamental uncertainty, but doing so in contexts where our conventional civic epistemology would most call for certainty. The problem of reimagining technology governance to account for untrustworthy expertise acquires new complexity when the technologies being governed have catastrophic hazards. A full exploration of these questions certainly lies beyond the scope of this chapter, but I will close by sketching some contours of how they might be approached.

Any viable approach to this problem must begin by bringing the boundary between regimes into clearer view: helping publics, policymakers, and even practitioners, recognize that it exists. This is not straightforward, however; in part because it is rarely evident (especially to lay observers) *why* the credibility of technical risk claims should falter in certain contexts. So it is, for example, that reactor reliability assessments resemble those of any other technology: they are crafted by the same types of experts, expressed in the same quantitative idioms, and supported by the same institutional forms of validation. Indeed, they exhibit greater expenditure of time, effort, and expertise than most other exercises in technological risk assessment. From a distance, then, there appears to be no obvious reason to treat them with greater skepticism.

This counterintuitiveness is compounded by an absence of clear signals that a technology has exceeded the bounds of conventional risk governance. As with quantum thresholds or relativistic horizons, there is no single moment when familiar risk heuristics break down, and few reliable signs that a system has entered a qualitatively different regime. We might expect the accident record to reveal when systems have crossed this threshold, but it rarely does. Historically, we have not attempted many types of systems with reliability demands that exceed what tests and models can verify, and those we have attempted we have mostly built in modest numbers. This means those systems rarely accumulate enough service data to test the safety claims made about them. Even decades of flawless performance reveal little about technologies with claimed mean-times-to-failure of a million years, in other words, so their safety can appear sound even when it falls far short of what is promised, as critics of nuclear deterrence infrastructures have long protested (Sagan 1993; Schlosser 2013). If we built a single doomsday device on the assurance it would catastrophically malfunction no more than once every billion years, it could be a million times less reliable than claimed and we likely wouldn't discover the discrepancy for centuries.

There is danger in this opacity; for when we cannot see that we are operating beyond the limits of our objectivity, our confidence becomes a source of risk in itself. In the absence of clear cues, we carry the old civic epistemology forward, treating implausible claims as if they were ordinary safety assurances. Just as juries are encouraged to read the reported accuracy of DNA tests as the probability of guilt, in other words, so policymakers are led to interpret reductive reliability calculations as meaningful measures of catastrophic risk. The result is a category mistake that hides the need for new standards of reasoning and oversight and sustains policies that mistake the appearance of safety for its reality.

The stakes are rising as systems that fracture our objectivity continue to proliferate. This is evident even in a simple tally of technologies with catastrophic failure hazards. Nuclear systems are exemplary in this respect. I have spoken at length about reactors, but they form part of a larger constellation of hazardous technologies, including spent-fuel stores, reprocessing facilities, and long-term waste repositories (Ramana 2018).[19] Extending outward in this constellation brings us to atomic weapons and the vast command-and-control infrastructures that govern their use. Failures in the latter have, at times, come perilously close to instigating thermonuclear war (Sagan 1993; Schlosser 2013).

Nuclear technologies are not alone in this regard, however rivaling them in catastrophic potential is a growing constellation of technologies that carry biological hazards. Engineered pathogens (Yassif et al. 2023), gene drives (Dolezel et al. 2020), and synthetic biology platforms (Zhang et al. 2024; Downer 2020) each carry the potential to trigger pandemics or irreversibly alter ecosystems. It is far from implausible that an accident at a biological research facility sparked the COVID-19 pandemic (Chan 2024). If the pandemic was, indeed, a lab leak, then it arguably represents the most calamitous technological failure in history, with a death toll in the millions.

Widening the lens further reveals a wide array of other technologies that harbor failure hazards severe enough to require extraordinary assurances. Chemical megaplants have the potential to release lethal toxins, such as those that caused the 1984 tragedy at Bhopal (Fortun 2001). Ports stockpile substances that can explode with enough force to destroy city blocks, as occurred at Tianjin in 2015 and Beirut in 2020 (Babrauskas 2016; El Sayed 2022). Ultra-deep drilling platforms operate under forces that test the limits of containment, as evidenced by the 2010 Deepwater

[19] Technically, it was a damaged spent fuel pool at Fukushima that caused the crisis that threatened Tokyo (Lochbaum et al. 2014).

Horizon disaster, which cost billions to control and polluted 1100 miles of coastline and 68,000 square miles of ocean (NCBP 2011). Breakdowns in grid-scale energy or fuel infrastructures could cascade into national crises (Zhao et al. 2025). The collapse of a global control system, such as a financial clearinghouse or cloud infrastructure, might be as destabilizing as any physical catastrophe (Perrow 2007). AI systems are increasingly being spoken of in similar terms, with experts fearing they could destabilize markets, empower malicious actors, or disrupt critical infrastructures in ways that spiral beyond human control and cause immense harms (Grey and Segerie 2025). And this is far from an exhaustive list.

It is reasonable to believe that none of our conventional governance frameworks is suited to managing the reliability demands of technologies with hazards at these scales. Engaging meaningfully with such systems, whether as scholars, policymakers, or simply as citizens, means reflecting more carefully and critically on the assertions experts make about their safety. Absent the pretense of objectivity, with its useful claim to finality, it becomes difficult to argue that such assertions should be delegated to a single expert authority. As with economic forecasts, it is normal in these circumstances to expect that experts will disagree about the "facts" of technological risks. And, in light of this, the goal becomes that of optimizing decision-making amid conditions of uncertainty and principled disagreement. This calls for practices that expose differing opinions and accommodate dissent. Instead of one unified expert authority, for instance, it makes more sense to imagine a system that cultivates multiple, independent experts and opinions, and allows for the coexistence of competing authorities, each with distinct interests.

Cultivating experts with distinct interests matters in this context because in situations where competing experts are making different interpretive choices, those choices are inevitably shaped by interests. Competing claims about risk must be read through the lens of who makes them, under what constraints, and to whose benefit. It becomes dangerous, for instance, to overlook whether those who attest to a system's safety have a stake in its acceptance. Here, the politics of knowledge that have long preoccupied STS scholars become directly relevant to the governance of risk, and Perrow (1999) becomes pertinent for its emphasis on structural incentives as key determinants of safety.

Perrow (1999) is similarly pertinent in this context for its emphasis on the inherent limits of safety, and the institutional consequences those limits entail. Much like the book's eponymous Normal Accidents, fractured objectivity [and the unavoidable "rational" or "epistemic" accidents it implies (Downer 2011, 2024)] forces us to reckon with the scope of our technological ambitions. For while our conventional civic epistemologies already allow that some technologies should be resisted on moral grounds—human cloning, for instance, or biological weapons—they rarely concede that desirable technologies should be avoided because their hazards exceed our control. Where we cannot wholly trust expert assurances about extreme hazards, however, restraint arguably becomes the only responsible course of action.[20] No

[20] It is not impossible to push the reliability of complex technologies beyond the epistemological limits of the tests and models by which they are known. The remarkable, and empirically undeniable,

level of capital expenditure, engineering rigor, or organizational sophistication will ever fully compensate for the fundamental epistemological limits of our knowledge practices; and those limits will always hide potential accidents (Downer 2011, 2024). We are free to choose our technological futures, but we need to own the consequences of those choices, and it is an error to believe we can build complex systems that simply cannot fail. Herein, perhaps, lies the most durable lodestar for navigating regimes of fractured objectivity (The fact that we *can* doesn't always mean that we *should*, Dr. Frankenstein).

As a practical matter, of course, such technologies cannot always be resisted; if only because, in a world of strategic competition and imperfect trust, there are risks to *not* building certain systems. Nuclear weapons and deterrence infrastructures are the cleanest examples of this: they may never be as safe as their hazards should require, but the risks of maintaining them must be weighed against those of unilateral disarmament (Sagan 1993).[21] A similar tension is emerging around advanced AI, where competition arguably compels states and firms to develop systems with risks they cannot wholly negate, lest they be outpaced and subsequently dominated by rivals with less restraint (Krakowski et al. 2023).

How states should navigate this tension, between the epistemic humility that fractured objectivity demands and the strategic imperatives that seem to foreclose it, is a question this chapter can only open. It implicates problems of democratic legitimacy, international coordination, and risk tolerance that deserve separate treatment. Here, I note only that such tensions bring us back to debates outlined earlier, about the merits of performing objectivity and managed displays of confidence.

It is tempting to say that where states feel compelled to build hazardous systems they cannot expect to fully control, they should at least seek legitimacy without the public pretense of objectivity and certainty. But on complex strategic questions—where the choices involved can be both highly technical and, simultaneously, too significant to be made in error—democracies arguably need to prioritize expertise over full transparency or direct consent (Schumpeter 2010 [1942]).[22] And in matters pertaining to technological risks, it is plausible that this might require meretricious performances of objectivity.

An underappreciated danger here, however, is that in performing objectivity for publics, states come to mislead themselves. This is to say that in focusing on "proving" that accidents cannot occur (i.e., risk), they forgo the prudence of planning for failures (i.e., resilience). Because why plan for contingencies that are not officially recognized as possible? Returning to reactors, for instance, we might note that

---

reliability of jetliners proves this. But pushing past these limits requires a protracted and disciplined process of painstakingly accumulating insights from failures, and, as such, it is not feasible in contexts where technologies have societally-intolerable failure hazards (Downer 2017, 2024).

[21] The politics of atomic energy might also be understood in this light. Refrains about new reactors being necessary for combatting climate change are misleading (Präger et al. 2024; Ramana 2024), but there is a stronger argument that they are necessary for sustaining atomic weapons. Stirling and Johnstone (2019), for instance, argue that the UK civilian nuclear power program is kept alive to maintain military nuclear capabilities.

[22] Brexit arguably stands as testament to this.

they are sometimes sited close to major population centers, and that neither evacuation plans nor medication stockpiles are ever commensurate with meltdowns (Kahn 2011; Clarke and Perrow 1996). The Fukushima accident revealed a spectrum of such shortcomings. For example, US nuclear authorities were unable to model its fallout beyond 50 miles (80 km), because their software for doing so was designed only for small leaks (Lochbaum et al. 2014, 63).[23]

The difficulties of mitigating this danger are often more political than technical. It is perfectly possible to stockpile medications or site reactors farther from major population centers. The challenge is institutional. Once authorities have formally certified a system as safe enough that such precautions are unnecessary, they bind themselves to that claim in budgets, liability frameworks, and public rhetoric. Measures that imply residual risk begin to look like admissions of failure in this context.

But this feels like a tractable problem, especially relative to the stakes involved, and one worth addressing. For even if our authorities must sometimes act beyond the limits of their technical competence, accepting more hazard than necessity demands remains a moral failure. Where states feel compelled to legitimize technologies with extreme hazards, they have an obligation to manage those hazards to the fullest extent possible. Denial does not cancel the duty of control.

Acknowledgments My thanks to Sveta Milyaeva, Alex Gantman, and Tom Osborne for their comments on earlier drafts of this chapter. Also to the Center for the Governance of Artificial Intelligence (GovAI), which was good enough to host me while I worked on some of it.

References

V. Babrauskas, Explosions of ammonium nitrate fertilizer in storage or transportation are preventable accidents. J. Hazard. Mater. **304**, 134–149 (2016)

U. Beck, *Risk Society: Towards a New Modernity* (Sage Publications, London, 1992)

N. Behling, M.C. Williams, S. Managi, Regulating Japan's nuclear power industry to achieve zero accidents. Energy Policy **127**, 308–319 (2019)

R. Blumenthal, J. Hawley, *Bipartisan framework for U.S. AI act* (2023). https://www.blumenthal.senate.gov/imo/media/doc/09072023bipartisanaiframework.pdf. Accessed 6 Sept 2024

A. Chan, Why the pandemic probably started in a lab in 5 key points. *New York Times* (2024). https://www.nytimes.com/interactive/2024/06/03/opinion/covid-lab-leak.html. Accessed 21 Oct 2025

L. Clarke, C. Perrow, Prosaic organizational failure. Am. Behav. Sci. **39**(8), 1040–1056 (1996)

S. Conrad, Z. Frank, S. Hurley, The Vantasner danger meridian (Season 2, Episode 2) [TV series episode], in S. Conrad (Executive Producer), *Patriot*. Amazon Studios (2018)

E. Constant, Reliable knowledge and unreliable stuff: on the practical role of unreliable beliefs. Technol. Cult. **40**(2), 324–357 (1999)

L. Daston, P. Galison, *Objectivity* (Zone Books, New York, 2007)

M. Dolezel, C. Lüthi, H. Gaugitsch, Beyond limits–the pitfalls of global gene drives for environmental risk assessment in the European Union. BioRisk **15**, 1–29 (2020)

[23] Assertions by US authorities about their models never showing any threat to the embassy in Tokyo need to be understood in this context.

J. Downer, When the chick hits the fan: representativeness and reproducibility in technological testing. Soc. Stud. Sci. **37**(1), 7–26 (2007)

J. Downer, '737-Cabriolet': the limits of knowledge and the sociology of inevitable failure. Am. J. Sociol. **117**(3), 725–762 (2011)

J. Downer, Disowning Fukushima: managing the credibility of nuclear reliability assessment in the wake of disaster. Regul. Govern. **8**(3), 287–309 (2014)

J. Downer, The aviation paradox: why we can 'know' jetliners but not reactors. Minerva **55**(2), 229–248 (2017)

J. Downer, On ignorance and apocalypse: a brief introduction to 'epistemic accidents,' in *New Perspectives on Safety Research*. ed. by J.C. Le Coze (CRC Press, 2020)

J. Downer, *Rational Accidents: Reckoning With Catastrophic Technologies* (MIT Press, Cambridge, MA, 2024)

J. Downer, M.V. Ramana, Empires built on sand: on reactor safety assessment and regulation. Regul. Govern. **15**(1), 1304–1325 (2021)

M.J. El Sayed, Beirut ammonium nitrate explosion: a man-made disaster in times of the COVID-19 pandemic. Disaster Med. Public Health Prep. **16**(3), 1203–1207 (2022)

K. Fortun, *Advocacy After Bhopal: Environmentalism, Disaster, New Global Orders* (University of Chicago Press, 2001)

Fukushima Nuclear Accident Independent Investigation Commission (NAIIC), *The Official Report of the Fukushima Nuclear Accident Independent Investigation Commission (Executive Summary)* (National Diet of Japan, Tokyo, 2012). http://www.nirs.org/fukushima/naiic_report.pdf. Accessed 16 July 2013

T.F. Gieryn, Boundary-work and the demarcation of science from non-science: strains and interests in professional ideologies of scientists. Am. Sociol. Rev. **48**(6), 781–795 (1983)

M. Grey, C.R. Segerie, The AI risk spectrum: from dangerous capabilities to existential threats. *arXiv preprint* arXiv:2508.13700 (2025)

S. Hodkinson, Grenfell foretold: a very neoliberal tragedy. Social Policy Rev. **30**, 5–26 (2018)

E. Hollnagel, *Safety-I and Safety-II: The Past and Future of Safety Management* (Ashgate, 2014)

IAEA (2005) Chernobyl: the true scale of the accident. Joint News Release WHO/IAEA/UNDP. https://inis.iaea.org/records/3p9y3-7aa33

S. Jasanoff, Technologies of humility: Citizen participation in governing science. Minerva **41**(3), 223–244 (2003)

S. Jasanoff, *Designs on Nature: Science and Democracy in Europe and the United States* (Princeton University Press, Princeton, 2005)

L.H. Kahn, Is the United States prepared for a nuclear reactor accident? Bull. Atom. Sci. (2011)

S. Krakowski, J. Luger, S. Raisch, Artificial intelligence and the changing sources of competitive advantage. Strateg. Manag. J. **44**(6), 1425–1452 (2023)

T. Kuhn, *The Structure of Scientific Revolutions*, 3rd edn. (University of Chicago Press, Chicago, (1996 [1962]))

B. Latour, *Science in Action: How to Follow Scientists and Engineers Through Society* (Harvard University Press, Cambridge, MA, 1987)

E.T. Layton Jr., American ideologies of science and engineering. Technol. Cult. **17**(4), 688–701 (1976)

D. Lochbaum, E. Lyman, S. Stranahan, *Fukushima: the Story of a Nuclear Disaster* (New Press, New York, 2014)

M. Lynch, S. Cole, *Truth Machine: The Contentious History of DNA Fingerprinting* (University of Chicago Press, Chicago, 2008)

D. MacKenzie, The Certainty Trough, in *Exploring Expertise: Issues and Perspectives*. ed. by R. Williams, W. Faulkner, J. Fleck (Macmillan, London, 1998), pp.325–329

A. Narayanan, S. Kapoor, AI existential risk probabilities are too unreliable to inform policy (2024). https://www.aisnakeoil.com/p/ai-existential-risk-probabilities. Accessed 9 Sept 2025

National Commission on the BP Deepwater Horizon Oil Spill and Offshore Drilling (NCBP), *Deep Water: The Gulf Oil Disaster and the Future of Offshore Drilling*. Report to the President (2011)

N. Oreskes, *Why Trust Science?* (Princeton University Press, Princeton, NJ, 2019)
C. Perrow, *Normal Accidents: Living with High-Risk Technologies*, 2nd edn. (Princeton University Press, Princeton, NJ, 1999)
C. Perrow, *The Next Catastrophe: Reducing Our Vulnerabilities to Natural, Industrial, and Terrorist Disasters* (Princeton University Press, Princeton, NJ, 2007)
H. Petroski, *To Engineer Is Human: The Role of Failure in Successful Design* (Vintage, New York, 1992)
G. Pinter, An Anatomy of the 2022 Gilt Market Crisis. *Bank of England. Staff Working Paper 1019* (2023). https://www.bankofengland.co.uk/-/media/boe/files/working-paper/2023/an-anatomy-of-the-2022-gilt-market-crisis.pdf. Accessed 26 Aug 2025
T. Porter, *Trust in Numbers: The Pursuit of Objectivity in Science and Public Life* (Princeton University Press, Princeton, NJ, 1995)
F. Präger, C. Breyer, H.J. Fell, C. Von Hirschhausen, C. Kemfert, B. Steigerwald et al., Evaluating nuclear power's suitability for climate change mitigation: technical risks, economic implications and incompatibility with renewable energy systems. Front. Environ. Econ. **3**, 1242818 (2024)
M.V. Ramana, Technical and social problems of nuclear waste. Wiley Interdiscip. Rev. Energy Environ. **7**(4), e289 (2018)
M.V. Ramana, *Nuclear is Not the Solution: The Folly of Atomic Power in the Age of Climate Change* (Verso Books, 2024)
M.V. Ramana, Beyond our imagination: Fukushima and the problem of assessing risk. Bull. Atom. Sci. (2011). https://thebulletin.org/2011/04/beyond-our-imagination-fukushima-and-the-problem-of-assessing-risk/. Accessed 30 Sept. 2025
A. Rip, The mutual dependence of risk research and political context. Sci. Technol. Stud. **4**(3/4), 3–15 (1986)
S. Sagan, *The Limits of Safety* (Princeton University Press, Princeton, NJ, 1993)
J.M. Samet, J. Seo, The financial costs of the Chernobyl nuclear power plant disaster. USC Institute of Global Health (2016). https://globalhealth.usc.edu/wp-content/uploads/2016/01/2016_chernobyl_costs_report.pdf. Accessed 28 Sept 2025
E. Schlosser, *Command and Control: Nuclear Weapons, the Damascus Accident, and the Illusion of Safety* (Allen Lane, New York, 2013)
J.A. Schumpeter, *Capitalism, Socialism and Democracy* (Routledge, London, (2010 [1942]))
S. Shapin, *A Social History of Truth: Civility and Science in Seventeenth-Century England* (University of Chicago Press, Chicago, 1994)
S. Silbey, Taming prometheus: talk about safety and culture. Ann. Rev. Sociol. **35**(1), 341–369 (2009)
A. Stirling, P. Johnstone, Are hidden military pressures for cross-subsidies driving major UK energy infrastructure decisions?, in *Written evidence submitted to the UK House of Commons BEIS Committee* (FEI0085) (2019). https://committees.parliament.uk/writtenevidence/101318/html/. Accessed 23 Oct 2025
W.C. Thompson, E.L. Schumann, Interpretation of statistical evidence in criminal trials: the prosecutor's fallacy and the defense attorney's fallacy. Law Hum Behav. **11**(3), 167–187 (1987)
T.R. Wellock, A figure of merit: quantifying the probability of a nuclear reactor accident. Technol. Cult. **58**(3), 678–721 (2017)
B. Wynne, Unruly technology: practical rules, impractical discourses and public understanding. Soc. Stud. Sci. **18**, 147–167 (1988)
A. Yablokov, V. Nesterenko, A. Nesterenko (eds.), Chernobyl: consequences of the catastrophe for people and the environment. Ann. New York Acad. Sci. **1181** (2009)
J.M. Yassif, S. Korol, A. Kane, Guarding against catastrophic biological risks: preventing state biological weapon development and use by shaping intentions. Health Secur. **21**(4), 258–265 (2023)
X. Zhang, T. Zhang, X. Wei, Z. Xiao, W. Zhang, Reducing potential dual-use risks in synthetic biology laboratory research: a dynamic model of analysis. Human. Social Sci. Commun. **11**(1), 1–14 (2024)

Y. Zhao, B. Cai, V. Cozzani, Y. Liu, Failure dependence and cascading failures: a literature review and research opportunities. Reliab. Eng. Syst. Saf. **256**, 110766 (2025)

GPSR Compliance

The European Union's (EU) General Product Safety Regulation (GPSR) is a set of rules that requires consumer products to be safe and our obligations to ensure this.

If you have any concerns about our products, you can contact us on ProductSafety@springernature.com

In case Publisher is established outside the EU, the EU authorized representative is:

Springer Nature Customer Service Center GmbH
Europaplatz 3
69115 Heidelberg, Germany

Batch number: 10791480

Printed by Printforce, the Netherlands